Animal Husbandry and
Livestock Management

Animal Husbandry and Livestock Management

Vincent Martin

R CALLISTO REFERENCE

www.callistoreference.com

Callisto Reference,
118-35 Queens Blvd., Suite 400,
Forest Hills, NY 11375, USA

Visit us on the World Wide Web at:
www.callistoreference.com

ISBN: 978-1-64116-228-9 (Hardback)

Cataloging-in-Publication Data

Animal husbandry and livestock management / Vincent Martin.
    p. cm.
Includes bibliographical references and index.
ISBN 978-1-64116-228-9
1. Animal culture. 2. Livestock. 3. Livestock--Management. I. Martin, Vincent.
SF61 .A55 2019
636--dc23

Table of Contents

Preface

Animal husbandry is the branch of agriculture that is concerned with the selective breeding, caring and raising of livestock for meat, milk, leather, eggs, etc. A wide variety of animals are raised for their products. Some of the commonly raised animals include cow, sheep, goat and pig. Species such as llama, rabbit, horse, guinea pig, etc. are also raised as livestock in some parts of the world. Modern animal husbandry focuses on intensive farming practices for meeting market demand. Animal breeding strives to ensure higher growth rate, low feed consumption for each unit of growth, prolificness, higher yields, etc. Artificial insemination and embryo transfer are common practices that are used today. This textbook is a compilation of chapters that discuss the most vital concepts in the field of animal husbandry and livestock management. It elucidates the concepts and innovative models around prospective developments with respect to raising livestock. This book is a complete source of knowledge on the present status of this important field.

Given below is the chapter wise description of the book:

Chapter 1- Animal husbandry is a branch of agriculture, concerned with animals raised for their meat, milk, fiber and eggs. Livestock refers to the domesticated animals that are reared for their produce. This chapter discusses in detail about livestock and animal husbandry, livestock farming and its various forms such as organic, sustainable and intensive livestock farming.

Chapter 2- Animal breeding is a branch of animal science, which deals with the evaluation of the genetic value of livestock. Animals with superior growth rate, production output, etc. are generally selected for breeding. The different types of animal breeding discussed in this chapter, include selective breeding, inbreeding, outbreeding and cross breeding. The topics elaborated in this chapter will help in developing a better perspective about the practice of animal breeding.

Chapter 3- Poultry farming refers to the rearing of poultry, which include birds like turkeys, chicken, ducks and geese for their meat and eggs. In this chapter, the significant concepts of poultry feeding, housing and various techniques of poultry management, such as debeaking and dubbing are explained in detail.

Chapter 4- Cattle are raised for meat, milk and hides. Some routine cattle husbandry practices are cleaning, feeding and milking. This chapter has been carefully written to provide an easy understanding of important cattle farming practices such as automatic milking, milk separation, use of milk cooling tanks and Bovine Somatotropin, etc.

Chapter 5- Horse breeding is a process of human-directed selective breeding of horses in order to achieve the desired characteristics in the horse. It is done particularly in purebred domesticated horses. Horse management refers to the attention and care of horses in order to ensure their optimal health and long life. This chapter addresses some of the basic practices of husbandry for horses and horse care, such as horse grooming, horse sheath cleaning, hoof care, the correct use of stable bandage, besides others.

Chapter 6- A wide range of species are reared as livestock. Some of the common livestock animals are sheep, goat, pig, etc. This chapter elucidates the different techniques and methods associated with the farming of the aforementioned livestock animals and their advantages. It has been carefully written to provide an easy understanding of the varied aspects of livestock farming.

Indeed, my job was extremely crucial and challenging as I had to ensure that every chapter is informative and structured in a student-friendly manner. I am thankful for the support provided by my family and colleagues during the completion of this book.

Vincent Martin

Chapter 1

Livestock Farming and Animal Husbandry

Animal husbandry is a branch of agriculture, concerned with animals raised for their meat, milk, fiber and eggs. Livestock refers to the domesticated animals that are reared for their produce. This chapter discusses in detail about livestock and animal husbandry, livestock farming and its various forms such as organic, sustainable and intensive livestock farming.

Livestock

Livestock are domesticated animals raised in an agricultural setting to produce commodities such as food, fiber, and labor. The term is often used to refer solely to those raised for food, and sometimes only farmed ruminants, such as cattle and goats.

OR

Animals kept or raised for use or pleasure; especially, farm animals kept for use and profit.

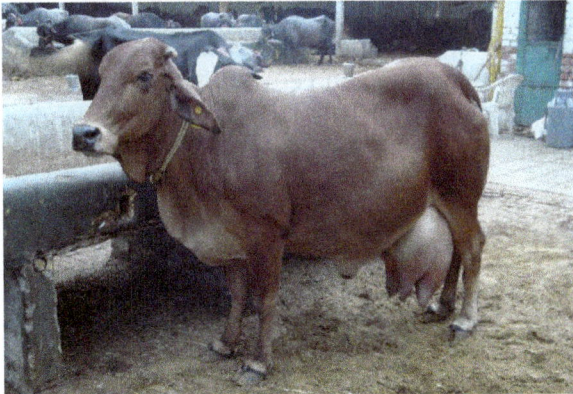

United States federal legislation sometimes more narrowly defines the term to make specified agricultural commodities either eligible or ineligible for a program or activity. For example, the Livestock Mandatory Reporting Act of 1999 defines livestock only as cattle, swine, and sheep. The 1988 disaster assistance legislation defined the term as "cattle, sheep, goats, swine, poultry (including egg-producing poultry), equine animals used for food or in the production of food, fish used for food, and other animals designated by the Secretary."

The word "livestock" is an umbrella term used for domesticated animals raised in an agricultural environment, with the intent of providing food, textiles, labor, or fertilizer to their owners. Common examples are horses, pigs, goats, cows, sheep, and poultry, although numerous other semi-wild animals including reindeer, yaks, camels, and emus could also be considered livestock. Humans have coexisted with domesticated animals for centuries, and the rise of farming and keeping animals probably contributed to a major shift in human culture.

The word can be taken to have several meanings, depending on interpretation. Livestock is sometimes referred to as "stock," in shorthand, reflecting the idea that the animals are property in addition to living beings. These animals are both living stock, or inventory, and the stock, or basis, of life for farmers and the people who rely on them. Raising animals is an important part of life for people all over the world.

Purely domesticated animals such as cows and horses are radically different than their wild counterparts. In some cases, the wild ancestors of domesticated livestock are actually extinct, because humans have selectively bred domesticated versions for so long. Domesticated livestock would probably have difficulty surviving in the wild, because it has been bred to be smaller and more docile than a wild animal would be. Semi-wild animals used as livestock, such as rabbits, have thriving wild populations in addition to domestic ones.

The uses for livestock are myriad. The most obvious is food, in the form of meat, dairy, and egg products. Few animals, however, are raised purely for their meat; the most notable exception to this is the pig, which is primarily a food animal. Most animals also contribute something else to the farm. Sheep, for example, have thick wool coats which are annually sheared to make textiles, and cows can provide physical labor as draft animals in addition to being a source of food. All livestock also produces plentiful amounts of manure, in the form of excrement, thus helping out in the farm garden as well.

Cattle in French Equatorial Africa

Some of these animals are also kept as pets, and enjoy privileged positions in human society. Horses, for example, are widely ridden and used as work animals, and in many cultures they have a status which borders on the sacred, while others have no difficulties eating their horses. In areas where living conditions are difficult, such as Tibet, a single livestock animal like the yak may provide the bulk of food, shelter, and companionship; therefore, the animals are highly valued.

Large animals such as horses, water buffalo, oxen and cattle are used in farming, particularly for plowing; are kept for wealth; and used as beasts of burden and sources of meat. Small livestock such as goats, pigs and sheep reproduce quickly and are sold for money, slaughtered for meat and raised for milk, skins and wool. Chickens, ducks and other fowl produce eggs and meat. Many of these animals are also used in sacrifices and rituals.

Villagers like to own their own plowing animals. It is difficult to borrow or share animals because often farmers want to plow and perform other chores with them at around the same time. Pigs, goats and chickens are generally allowed to roam free around the village. They sometimes roam around inside huts where people live. Development officials try to discourage this as a precaution against disease.

Modern domesticated animals are the way the are because of selecting breeding, which essentially means to take animals with traits you want and breed them and in the next generation you repeat the process until the traits become dominate.

Diversified uses of Livestock

Domestic animals have, for more than 10 thousand years, contributed to human needs for food and agricultural products. These products include meat, dairy products, eggs, fibre and leather, draft power and transport, and manure to fertilize crops and for fuel. These animals have always played a large cultural role for livestock keepers. Livestock also play an important economic role as capital and for social security.

The value of livestock has also been clearly demonstrated for soil nutrient management, especially in soils in rapidly intensifying crop-livestock systems and in those already intensified. Integration of livestock into crop systems enhances smallholder farm productivity and profitability.

The multiple uses of livestock also include their cultural roles in many societies. Consequently, the use of animal resources varies considerably in different parts of the world, as the social, environmental and other conditions for animal production differ enormously.

Currently, an estimated 30-40% of the world's total agricultural output is produced by its variety of livestock. In some parts of the world, including some parts of Africa where intensive mixed livestock-crop systems are practised, as much as 70-80% of the farm income is from livestock. In such systems, much of the crops produced are fed to livestock and converted to high quality food for human consumption.

Adaptation to Environment a Necessity

In most parts of the developing world, difficult environmental conditions and a lack of availability of capital, technology, infrastructure and human resources have not allowed intensification of agriculture, including development of genetic resources. Instead, harsh climate, little feed of low nutritional value, irregular feed availability, diseases, and lack of education and infrastructure, have kept the agricultural output per animal at a low and rather unchanged level for a long time. However, livestock breeds in the tropical parts of the world have during thousands of years become adapted to cope with harsh environments, including disease challenges, and to produce under conditions in which breeds developed in more favourable environments will not even survive. Such differences among animal populations have a genetic background and are the result of the interaction between genetic constitution and environment. This has evolved over time from natural and human selection of animals for performance in different environments. That is why there is such a variety of indigenous breeds. However, when appropriately utilized in pure or cross-breeding programmes, indigenous breeds can contribute to increased productivity in smallholder production systems.

Increased Productivity to Avoid Degradation of Natural Resources

The challenge now is to find ways to exploit the potential for improved and sustainable livestock production that the variability among and within the indigenous breeds may offer different environments and production systems in various parts of the tropics and sub-tropics. Otherwise, it will not be possible to produce what is needed for the people of the developing world to survive. To date, demand for increased livestock production has largely been met by increasing the number of indigenous animals without improving yield or efficiency per animal or area used. Such trends will not hold in future as industrialization is predicted to continue at a higher pace, especially for pig and poultry production, using mainly genetically improved breeds and composites. Non-structured cross-breeding of indigenous breeds with imported high yielding breeds has been practised too often in the tropics, sometimes with disastrous results. This development cannot continue.

Land degradation and the increasing amount of resources required to just maintain the animal populations must be replaced by more efficient systems demanding higher outputs per animal or area of land used to meet the future demands of livestock products. For sustainability, these systems must emphasize effective resource input/output ratios and more integration of livestock and crop production rather than industrialized mono-cultural production systems that seriously challenge the wise use and care of our natural resources.

Consumer concern and consumer perceptions in light of the increasing global push for product standardization and wider impacts of production systems on environments are of increasing concern. Whereas such trends provide potential scope for environmentally friendly produced livestock products, the effects of over-exploitation (deforestation and overgrazing) of common and open access resources, especially by the rural poor, may undermine the potential gains. Besides, to fully benefit from better prices offered by niche markets for more naturally produced products, better levels of producer organization, in terms of product quality assurance, standardization and general marketing, will be required of producers to enable such potentials to be exploited.

It is rightly argued that animal production systems, especially with ruminants, contribute to undesired methane emissions. However, it is also well established that these greenhouse emissions can be substantially reduced by increasing productivity and lowering the number of animals kept for a given total amount of produce. Hence, increased productivity per animal concentrating production on fewer but more valuable animals is a way forward in reducing the negative environmental impacts of livestock production. This intensification must, however, also be designed to effectively manage all other risks to environmental degradation of land and water, e.g. efficient ways of using manure and wastes from other farm products. For example, in large commercial tree plantation systems such as those in Malaysia, increased resource utilization and profitability may arise from integration of livestock in rubber and palm oil plantations. Such integration also has the potential for reducing the country's annual demands for imported beef and milk to meet the domestic deficits.

More productive breeds of a number of livestock species have been genetically developed to fit different markets and environments for both developed and developing countries. Such genetic changes, in combination with better and continuously available feeds and management, have in a few decades led to the doubling of food production in a number of breeds and species. Such increases in agricultural produce require high technology and large inputs of feed, labour, energy and capital, and good disease control and management practices. However, in high input and resourceful industrialized systems, limited considerations regarding total efficiency in nutrient cycling and pollution have been made. Without such considerations, these production systems will not be sustainable. Conversely, in low and medium input pasture production systems small ruminants, camels and beef cattle provide the most efficient way of utilizing such environments to produce valuable livestock products (milk, meat and leather). To date, the potentials of many of the indigenous livestock populations and breeds remain largely unexploited. Through well organized conventional selection programmes much more could be achieved. Exploitation of local and foreign niche markets that favour the smaller and more adapted indigenous breeds exist in the Middle East and in many Asian countries. Strategic use of such breeds as dam-lines/breeds in terminal cross-breeding programmes presents great potential and prospects.

Most local breeds are kept under smallholder systems, though pastoralists may also keep large herds. The role of the smallholder farmers may also be important in the future, but most likely the production will need to be intensified. Smallholder animal production may need to be combined with crop production, and be relocated to peri-urban and urban areas. This will require increasing focus on environmental and product quality issues and on market access and competitiveness. The interaction between genotypes and environments would continue to be a key element in choice and development of future breeding stocks while some environmental changes, such as improved feeding and management practices, will also have to take place.

Animal Husbandry

Animal husbandry is the science of farming of animal livestock. It includes caring, breeding and management of livestock. Animal husbandry is a large scale business where animals that provide us food are reared, bred, sheltered and cared in a farm or regions which are specially built for them. Animal husbandry was initiated with cattle farming. Under the cattle farming, livestock such as cows, goats, buffalo, sheep, etc. are reared. Later, animal husbandry was even extended to poultry farming, fisheries, apiculture, etc. And this extends a helping hand to the increasing needs of the generations.

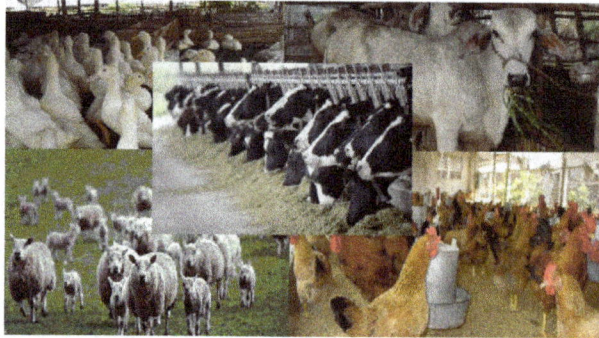

Animal husbandry, agriculture and allied activities have been the core livelihood for majority of the rural people since time immemorial. It provides productive employment, especially self employment and the most valuable supplementary income to a vast majority of rural households, majority of who are small and marginal farmers and landless laborers. Livestock provide increased economic stability to the poor masses. They act as cash buffer in case of small stock and as captive reserve in case of larger stock. Livestock provide quality animal protein to human population in the form of milk, eggs, meat and value added products. They provide draught power for agricultural operations, organic manure for agriculture and raw materials like skin, hides, blood, bone, hoof, horn, etc., for various industries.

During 2006-07, the gross value of output of livestock in the State is Rs.965.43 crores. The contribution of livestock sector to the Gross State Domestic Product is 3% and that to the agriculture and allied activities is 28%. The dairy and poultry sectors contributed significantly to this growth. During 2005-06, the estimated milk and egg production in the State was 54.74 lakhs MT and 6,223 millions respectively. Likewise, the per capita availability of milk/day and eggs/annum was 234 gms and 97 nos. respectively. The State contributes 5% of milk and 14% of egg production and stands 8th in milk production and 2nd in egg production in the country.

Animal husbandry is one sector which has high potential for growth. The potential of the sector needs to be exploited as this can play a key role in providing sustainable employment in their location itself and arrest migration of people to urban areas. As animal husbandry is an activity which can easily be taken up by rural communities as skill and resource requirements are minimal, inputs are locally available and marketing does not pose a major problem, it can act as an engine in poverty alleviation programmes by making asset less poor into income generating asset owing population. This will go a long way in not only augmenting food security, human security, empowerment of women and rural youths, but will also help in triggering and invigorating the rural economy ultimately contributing significantly to the comprehensive socio-economic transformation of the state.

Cattle are domesticated ungulates, a member of the subfamily Bovinae of the family Bovidae. They are raised as livestock for meat (called beef and veal), dairy products (milk), and leather and as draught animals (pulling carts, ploughing etc). The Indian subcontinent harbours a variety of cattle. Besides many non-descript breeds, there are 30 well-recognized cattle breeds in India. Majority of these breeds are low producers of milk; hence they are primarily used for the production of castrated bulls, which are used in agriculture, carting and transport.

The total livestock population of the state is 249.41 lakh which is 5.07% of the country's livestock population. The state accounts for 4.94% cattle, 1.69% buffaloes, 9.105 sheep and 6.58% goats to the country.

According to the 17th Livestock and poultry census, the spices wise distribution of livestock and poultry, their percentage to the total and density in the state is follows.

Species	Population (in lakhs)	Percentage to total livestock	Density per sq.km.
Cattle	91.41	36.65	70
Buffalo	16.58	6.65	13
Sheep	55.93	22.42	43
Goat	81.77	32.79	63
Pig	3.21	1.29	2
Donkeys, Mules & Camels	0.51	0.20	-
Total Livestock	249.41	100	191
Dog	27.17	-	21
Poultry	865.91	-	666

Livestock Farming

Livestock farming is the rearing of animals for food and for other human uses. The word 'Livestock' applies primarily to cattle or dairy cows, chickens, goats, pigs, horses and sheep. Today, even animals like donkeys, mules, rabbits and insects such as bees are being raised as part of livestock farming.

The usefulness of livestock organs in medicines like insulin has been understood only recently. Nevertheless, livestock farms have been benefiting us in many ways for ages - they provide us with eggs, honey, meat, milk, etc. The skins or hides and even hair of these animals have been used to make blankets, clothing, shoes and the like. The hoofs and horns of these farm animals have been used to make common items like buttons and combs. You cannot ignore the specialty item, the bullhorn showpiece in your drawing room. Even the animal-wastes do not go waste - they make excellent natural fertilizers.

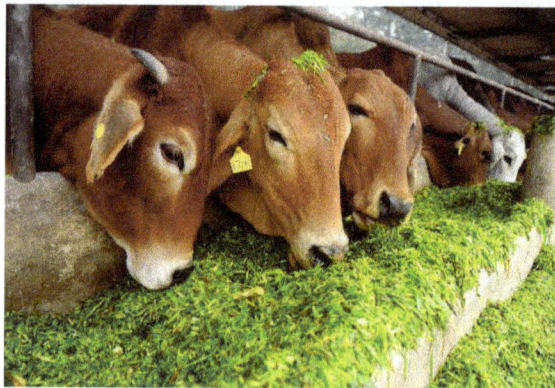

Present-day livestock farming is very well planned - animals are provided with food and shelter and bred selectively. As regards shelter, animals are usually kept in enclosures. Pigs and poultry are reared intensively in indoor environments. However, indoor animal farming has often been criticized for multiple reasons - on grounds of pollution and for animal welfare reasons. Outdoor farming of livestock stands for rearing animals in bigger enclosures like ranches and fenced pastures.

When it comes to feeds, farm animals are allowed to graze; additionally, they are given nutritious processed feed to aid normal growth and production.

The recent breeding practices that have altogether changed the scenario are:

- Selective Breeding: mating the best males and females, both related or unrelated.

- Artificial Insemination: artificially impregnating a female animal with semen from a superior-quality male.

- Embryo Transfer Techniques: where a genetically enhanced fertilized egg is transferred from the reproductive tract of a special breed cow into the reproductive tract of a second cow.

Consequently, we now have excellent breeds of farm animals that yield more and the produce-quality too is superior.

There have been huge improvements in livestock farming in the recent past, with each of them being more scientific aiding in improvement of animal health and increased yield.

Livestock farming is a double-benefiting system - it helps the farmers both economically and supplies them with food throughout the year. However, the above benefits do surely rest upon a two faced criterion - the development must proceed scientifically, and protection and care should be first on the farmer's list.

Organic Livestock Farming

According to the Small Farms and Alternative Enterprises department at the University of Florida:

"Organic livestock production often necessitates the integration of animal-pasture-crop production to be successful."

Producing organic livestock means that not only have you raised the animal on a continuous organic program from birth/hatching, their food must be of organic origin and they must also live in surroundings that allow for natural behavior and health. For example, confined feedlots for cattle or sheep are not natural. In addition, farmers must manage manure and waste products a way that recycles nutrients yet avoids contaminating water, soil or crops.

SFAE advises that producers cannot provide prophylactic antibiotics, medicines given to ward off sickness. Rather, they are encouraged to "treat animals with appropriate treatment, including antibiotics and other conventional medicines when needed, but treated animals cannot be sold or labeled as organic." And obviously, hormones or other growth drugs are taboo, as well.

Clearly, with these stipulations, keeping records is important, and as it turns out, the Organic Food Production Act dictates that farmers must maintain records for five years to verify the organic status of:

- animals
- production
- harvesting
- product-handling practices

Furthermore, official certification is required in order to sell products under an organic label. A quick internet search will send you in the proper direction, giving an overview for your state and contact information for the official agency in charge of organic certification. Save yourself a lot of potential financial and emotional stress—if you don't want to go through the time and expense of certification, don't put "organic" on your label.

For livestock like these healthy cows vaccines play an important part in animal health since antibiotic therapy is prohibited in organic farming

Raising livestock and poultry, for meat, dairy and eggs, is another traditional farming activity that complements growing. Organic farms attempt to provide animals with natural living conditions and feed. Organic certification verifies that livestock are raised according to the USDA organic regulations throughout their lives. These regulations include the requirement that all animal feed must be certified organic.

Organic livestock may be, and must be, treated with medicine when they are sick, but drugs cannot be used to promote growth, their feed must be organic, and they must be pastured.

Also, horses and cattle were once a basic farm feature that provided labor, for hauling and plowing, fertility, through recycling of manure, and fuel, in the form of food for farmers and other animals. While today, small growing operations often do not include livestock, domesticated animals are a desirable part of the organic farming equation, especially for true sustainability, the ability of a farm to function as a self-renewing unit.

Sustainable Livestock Farming

At first glance, sustainable livestock farming looks a whole lot like organic livestock farming; however, it is more like the umbrella under which we find organic practices. Sustainability takes in a much larger scope. The clue is in the name - Sustainable practices are those that can be sustained over time, as all parts contribute to the whole. Just a few examples of this symbiosis include no-till farming, preventing erosion and conserving water, while organically raised animals provide organic fertilizer for those no-till crops and help control weeds and pasturage. According to Oregon State University, sustainable livestock farming takes into account financial, environmental, ethical, social, product-quality and animal-welfare issues.

Sustainability can be overwhelming at first because it can extend past what we produce and how we produce it to address what we eat, the products we clean with, how we recycle and reuse our resources, and more. Our best bet, as hobby farmers desiring to employ sustainable livestock farming practices, is to start small. One change at a time is, well, more sustainable.

Healthy Soils and Water

Keeping permanent pasture in place and managing it with intensive rotational grazing is one of the best forms of agriculture known to prevent soil erosion and build healthy soils. Each group of animals are moved every 24 hours in the summer to a fresh paddock where animals are concentrated using portable electric fencing - polywire for the cows and Premier's Electronet for the sheep. Paddocks are grazed to 3-4 inches then allowed to "rest" for approximately 3 weeks before the next group grazes again when it is recovered to 8-10 inches in height. The animals are then not allowed to graze too low, the soil surface remains covered and is kept moist and cool so microorganisms and macro fauna stay healthy and happy. Also, by allowing the field to "rest", manure can breakdown and roots can grow deep into the soil again, improving nutrient cycling. This practice also protects water quality - when erosion is avoided, siltation of waterways does not occur and by keeping nutrients cycling in the pasture, those nutrients do not runoff.

Resilience

A wide range of plants in pastures is utilized, including orchard grass, perennial rye, meadow brome, timothy, white clover, red clover, alsike, birdsfoot trefoil, alfalfa, and chicory. Each of these plants has its own set of qualities - some fix nitrogen, some are deep rooted and therefore drought resistant, some do well in wet years, and all have their own nutrient profile, providing a balanced diet for the sheep and cows grazing them. Having a range of plants makes the pastures more resilient to climate extremes so no matter what the weather is like each year, there is always something growing well, both for the livestock but also for the soil microorganisms that need living plants to provide carbon and nitrogen into the soil profile.

Biodiversity and Parasite Management

Using a variety of plants is only one of the ways used biodiversity to improve the health and sustainability of the farm. In fact, the reason the cows were introduced to the system back in 2004 was to help do a better job of managing internal parasites in the sheep flock. "Worms" in sheep are a major problem for most shepherds because of the over-use of anthelmintics, or wormers, in modern agriculture. Parasites have developed resistance to most of the drugs on the market. Additionally, a healthy, sustainable livestock system should not have to rely on drugs to manage a common problem. A range of practices to manage parasites are used: Cows and sheep on each paddock are alternated as they rotate through the farm because they do not share the same species of parasites and by alternating, the parasite load on the pastures is reduced to a level that the flock can handle. Sheep that have natural resistance to parasites are also selected. The original flock was a mix of Finn, Dorset and East Friesian breeds but pure Icelandic rams have been in use for many years now and the percentage of Icelandic genetics in the flock have been constantly increased because they show a good tolerance of parasites. Finally, plants in the pasture mix that contain condensed tannins (especially chicory and birdsfoot trefoil), which have been shown to reduce the rate of infection by parasites in the sheep are included.

Humane Treatment

A number of practices endorsed by "Animal Welfare Approved" are followed - lamb on pasture in early May are used, the tails on sheep or cows are not docked, and every paddock is provided with shade and fresh water.

Greenhouse Gas Emissions

Managing for climate change mitigation and adaptation is a priority on farms. The fossil fuel energy usage on the farm is low but not zero - diesel is used tractors and the pastures are clipped 2-3 times a year to manage unwanted weeds. Hay for winter which are made using tractors are also bought.

Benefits of Sustainable Livestock Farming

Truly sustainable livestock farming requires the use of a pasture-based system. Pasture-raised animals roam freely in their natural environment where they are able to eat nutritious grasses and other plants that their bodies are adapted to digest. In addition to dramatically improving the welfare of farm animals, pasturing also helps reduce environmental damage, and yields meat, eggs, and dairy products that are tastier and more nutritious than foods produced on factory farms.

Animal Health Benefits

Animals raised on pasture enjoy a much higher quality of life than those confined within factory farms. When raised on open pasture, animals are able to move around freely and carry out their natural behaviors. This lifestyle is impossible to achieve on industrial farms, where thousands of animals are crowded into confined facilities, often without access to fresh air or sunlight. These stressful conditions are a breeding ground for bacteria and the animals frequently become ill, so factory farms must routinely treat them with antibiotics to prevent outbreaks of disease.

Grazing on pasture is especially beneficial for cattle and other ruminants, whose bodies are developed to eat grass. The roughage provided by grasses and other plants allows ruminants to produce saliva, which helps neutralize acids that exist naturally in their digestive systems. When taken off pasture and put on a diet of grain, a ruminant will produce less saliva, causing an increase in acidity within its digestive tract. As a result, grain-fed cattle often suffer from a number of health problems including intestinal damage, dehydration, liver abscesses and even death. Despite the fact that grain diets can sicken cattle and other ruminants, factory farms feed these animals grain (usually corn or soybeans) because it's a cheap way to fatten animals and force them to grow to market weight as quickly as possible.

Pasture-raised animals also enjoy a diet free of the unnatural feed additives routinely administered on factory farms. Industrial farms frequently supplement animal feed with a range of byproducts including chicken manure, plate waste from restaurants, and animal blood in order to bolster the quantity and protein content of the feed. Antibiotics and artificial hormones are also added to promote rapid growth. On pasture, animals get all the nutrients they need from grass and forage (other plants), and some animals, like chickens, get additional vitamins and protein from eating insects.

Human Health Benefits

A growing body of research indicates that pasture-raised meat, eggs, and dairy products are better for consumers' health than conventionally-raised, grain-fed foods. In addition to being lower in calories and total fat, pasture-raised foods have higher levels of vitamins, and a healthier balance of omega-3 and omega-6 fats than conventional meat and dairy products.

Studies have shown that milk from pasture-fed cows has as much as five times the CLA (a type of fatty acid) as milk from grain-fed cows. And meat from pasture-fed cows has from 200 to 500 percent more CLA as a proportion of total fatty acids than meat from cows that eat a primarily grain-based diet.

Free-range chickens have 21% less total fat, 30% less saturated fat and 28% fewer calories than their factory-farmed counterparts. Eggs from poultry raised on pasture have 10% less fat, 40% more vitamin A and 400% more omega-3's.

Environmental Benefits

Pasture-based systems can help the environment, especially through fertilizing the soil and by reducing the amount of grain produced as feed. And unlike industrial farms, which rely on large amounts of fossil fuels to truck feed and animal waste, pasture-based systems take advantage of the animal's ability to feed itself and spread its own manure.

Intensive Livestock Farming

Intensive livestock farming, referred to by opponents as "factory farming" or "industrial livestock production," is the name given to operations involving large numbers of animals that are being raised on a small amount of land. While these could include confined animal feeding set-ups, like feedlots for cattle, confined dairy herds or chickens grown at extreme densities, there are viable, sustainable ways to support a large number of animals without such situations.

Some farmers opt for intensive rotational grazing. Included in this might be growing meat chickens in chicken tractors. These movable pens are rotated over pasture as soon as the current spot becomes eaten down. Sheep, cattle and goats can be managed similarly, rotating them out of one paddock and into another at the optimal time that allows the pasture to recuperate quickly while affording fresh graze at all times.

Advantages and Disadvantages of Intensive Farming

Intensive farming is the latest technique used to yield high productivity by keeping large number of livestock indoors and using excessive amount of chemical fertilizers on a tiny acreage. It is carried out to meet the rising demand for cheap food and prevent future shortages. Elaborated below are its pros and cons. Intensive farming is an agricultural system that aims to get maximum yield from the available land. This farming technique is also applied in supplying livestock. You could say that under this technique, food is produced in large quantities with the help of chemical fertilizers and pesticides that are appropriately used to save such agricultural land from pests and crop diseases.

Products, such as eggs, meat, and other agricultural items that are easily available in many supermarkets today are produced using modern intensive farming methods. Jay Rayner, a restaurant critic of the Observer says, Sure, it might be cruel, but intensive farming saves lives. It is practiced widely by many developed economies of the world. Sustainable intensive farming, intensive aquaculture, intensive livestock farming, and management-intensive grazing fall under this farming category.

Advantages

- One of the major advantages of this farming technique is that the crop yield is high.

- It helps the farmer to easily supervise and monitor the land and protect his livestock from being hurt or hounded by dangerous wild animals.

- With the introduction of intensive farming, farm produce, such as vegetables, fruits, and poultry products have become less expensive. It also aids in solving the worldwide hunger problems to a great extent. This means that common people can now afford a balanced and nutritious diet.

- Many opine that organic food can be afforded only by the elite strata of the society. Apart from that, large farming spaces are required to cultivate organic crops using natural manure. However, with the introduction of intensive farming, the space, equipment, and other requirements for farming are less and more economical.

- The EPA (Environment Protection Agency) has set certain rules and regulations on how livestock, pesticides, and animal manure are to be maintained. The farmers, who follow these set rules help to provide an affordable, safe, and healthy produce to all alike.

- Another advantage is that large productivity of food is possible with less amount of land. This leads to economies of scale and directly contributes towards meeting the ever-growing demand for food supplies.

Disadvantages

- Intensive farming involves the use of various kinds of chemical fertilizers, pesticides, and insecticides. Apart from this, it is also associated with farms that keep livestock above their holding capacity, which in turn leads to pollution, various diseases, and infections brought about by overcrowding and poor hygiene.

- Reports and studies reveal that intensive farming affects and alters the environment in multiple ways. Forests are destroyed to create large open fields, and this could lead to soil erosion. It affects the natural habitat of wild animals. Use of chemical fertilizers contaminates soil and water bodies, such as lakes and rivers.

- Pesticides sprayed on crops not only destroy pests and contaminate the crops, but also kill beneficial insects. Heavy use of pesticides and chemical fertilizers also affects the workers (who spray the pesticides) and the people residing nearby. Eventually, these chemicals are passed on to human beings, who consume the agricultural produce.

- Fruits and vegetables purchased from farms that promote intensive farming are covered with invisible pesticides. These cannot be washed off easily. Exceeding the use of pesticides affects the health of human beings severely, leading to skin allergy, physical deformity, and congenital disease.

- Statistics show a direct relation between the consumption of food procured from intensive farming sites and an increase in the number of cancer patients and children born with defects. Researchers opine that consumption of inorganic poisonous vegetables, fruits, poultry, and meat could probably be one of the reasons for causing such damage in the human body.

- There are many hybrid varieties of livestock, plants, and poultry available today. The livestock and poultry are injected with hormones and other chemicals to increase the yield.

References

- Coleman, Eliot (1995), The New Organic Grower: A Master's Manual of Tools and Techniques for the Home and Market Gardener(2nd ed.), pp. 65, 108, ISBN 978-0930031756

- Watson CA, Atkinson D, Gosling P, Jackson LR, Rayns FW (2002). "Managing soil fertility in organic farming systems". Soil Use and Management. 18: 239–247. doi:10.1111/j.1475-2743.2002.tb00265.x. Preprint with free full-text

- Brinton W, et al. (2004). "Compost teas: Microbial hygiene and quality in relation to method of preparation" (PDF). Biodynamics: 36–45. Retrieved 15 April 2009

- Paull, John (2017) "Four New Strategies to Grow the Organic Agriculture Sector", Agrofor International Journal, 2(3):61-70

- Horne, Paul Anthony (2008). Integrated pest management for crops and pastures. CSIRO Publishing. p. 2. ISBN 978-0-643-09257-0

- Fließbach, A.; Oberholzer, H.; Gunst, L.; Mäder, P. (2006). "Soil organic matter and biological soil quality indicators after 21 years of organic and conventional farming". Agriculture, Ecosystems and Environment. 118: 273–284. doi:10.1016/j.agee.2006.05.022

- Dagget, Dan. "Convincing Evidence". Man in Nature. Archived from the original on 6 March 2001. Retrieved 5 April 2013

- Lotter, D. (2003). "Organic Agriculture" (PDF). Journal of Sustainable Agriculture. 21 (4): 59. doi:10.1300/J064v21n04_06

- Vogt G (2007). Lockeretz W, ed. Chapter 1: The Origins of Organic Farming. Organic Farming: An International History. CABI Publishing. pp. 9–30. ISBN 9780851998336

- Kremer, Robert J.; Li, Jianmei (2003). "Developing weed-suppressive soils through improved soil quality management". Soil & Tillage Research. 72: 193–202. doi:10.1016/s0167-1987(03)00088-6

- "European organic farming research projects". Organic Research. Archived from the original on 10 January 2014. Retrieved 10 January 2014

- Does organic farming reduce environmental impacts? - A meta-analysis of European research, H.L. Tuomisto, I.D. Hodge, P. Riordan & D.W. Macdonald, Authors' version of the paper published in: Journal of Environmental Management 112 (2012) 309-320

- Kirchmann, Holger; Bergstrom, Lars (16 December 2008). Organic Crop Production - Ambitions and Limitations. Springer Science & Business Media. pp. 2–. ISBN 978-1-4020-9316-6

Chapter 2

Animal Breeding

Animal breeding is a branch of animal science, which deals with the evaluation of the genetic value of livestock. Animals with superior growth rate, production output, etc. are generally selected for breeding. The different types of animal breeding discussed in this chapter, include selective breeding, inbreeding, outbreeding and cross breeding. The topics elaborated in this chapter will help in developing a better perspective about the practice of animal breeding.

A group of animals related by descent and similar in most characters like general appearance, features, size, configuration, etc. are said to belong to a 'breed'.

Animal breeding is producing improved breeds of domesticated animals by improving their genotypes through selective mating.

Objectives of Animal Breeding

The main objectives of animal breeding are:

    (i) improved growth rate,

   (ii) increased production of milk, meat, egg, wool, etc.,

  (iii) superior quality of milk, meat, eggs, wool, etc.,

  (iv) improved resistance to various diseases,

(v) increased productive life, and

(vi) increased or, at least, acceptable reproduction rate.

Methods of Animal Breeding

Two methods of animals breeding are: inbreeding and out breeding, based mainly on breeding work with cattle.

1. Inbreeding

When breeding is between animals of the same breed for 4-6 generations, it is called inbreeding. Inbreeding may be explained by taking an example of cows and bulls. Superior cows and superior bulls of the same breed are identified and mated. The progeny obtained from such mating are evaluated and superior males and females are identified for further mating. A superior female, in the case of cattle, is the cow that produces more milk per lactation.

On the other hand, a superior male is that bull, which gives rise to superior progeny as compared to those of other males. As the homozygous pure lines developed by Mendel as described in Chapter 5, a similar strategy is used for developing pure lines in cattle as was used in case of peas. Inbreeding, as a rule, increases homozygosis.

Thus inbreeding is necessary if we want to develop a pure line in any animal. Inbreeding exposes harmful recessive genes that are eliminated by selection. It also helps in accumulation of superior genes and elimination of less desirable genes. But continued inbreeding reduces fertility and even productivity.

This is called inbreeding depression. In this condition, the selected animals of the breeding population should be mated with superior animals of the same breed but unrelated to the breeding population. This often helps in restoring fertility and yield.

2. Out breeding

Out breeding is the breeding between the unrelated animals which may be between individuals of the same breed (but having no common ancestors) or between different breeds (cross breeding) or different species (interspecific hybridization).

(i) Out crossing:

It is the mating of animals within the same breed but having no common ancestors on either side of their pedigree up to 4-6 generations. The offspring of such a cross is called as an outcross. Outcrossing is the best breeding method for animals that are below average in productivity in milk production, growth rate in beef cattle, etc. Sometimes only one outcross helps to overcome inbreeding depression.

(ii) Cross-breeding:

In cross-breeding superior males of one breed are mated with superior females of another breed. Many new animal breeds have been developed by this strategy. It gives better breeds. Cows of an inferior breed may be mated to bulls of a superior breed to get better progeny. Hisardale is a new breed of sheep developed in Punjab by crossing Bikaneri ewes and Marino rams.

(iii) Interspecific Hybridisation:

In this approach, male and female animals of two different species are mated. The progeny obtained from such a mating are usually different from both the parental species.

But in some cases, the progeny may combine desirable characters of both the parents. Mule is produced from a cross between female horse (mare) and male donkey. Mules are harder than their parents and are well suited for hard work in mountainous regions.

Controlled Breeding Experiments

These are carried out using artificial insemination and Multiple Ovulation Embryo Transfer Technology (MOET).

(i) Artificial Insemination (AI):

The semen of superior male is collected and injected into the reproductive tract of the selected female by the breeder. The semen can be used immediately or can be frozen for later use. When a bull inseminates a cow naturally approximately 5 to 10 billion sperms are deposited in the vagina. However, when semen is deposited artificially, considerably fewer sperms are required to achieve conception. Therefore, artificial insemination is very economical. The spread of certain diseases can be controlled by this method.

(ii) Multiple Ovulation Embryo Transfer Technology (MOET):

In this method, hormones (with FSH-like activity) is given to the cow for inducing follicular maturation and super ovulation instead of one egg, which they usually give per cycle, they produce 6-8 eggs. The cow is either mated with a best bull or artificially inseminated.

The embryos at 8-32 cell stage are recovered and transferred to surrogate mothers. The genetic mother is available for another super ovulation. MOET has been done in cattle, sheep, rabbits, buffaloes, mares, etc. High milk giving breeds of females and high quality (lean meat with less lipid) meat- giving bulls have been bred successfully to obtain better breed in a short time.

Selective Breeding

Natural selection and selective breeding can both produce changes in animals and plants. The difference between the two is that natural selection occurs in nature, but selective breeding only occurs when humans intervene.

Selective breeding is a process where we choose the characteristics we want in an animal. We then breed together a male and female that show some of those characteristics. From the offspring produced we select those that show the characteristic the most, and breed them together.

This process is repeated over many generations, each time selecting and breeding together those animals that have the characteristics we are looking for. Over a large number of generations, this can produce some surprising results.

The procedure involves identifying certain desirable features and finding two members of a species that exhibit the particular feature. A series of matings or breedings is then performed between the individuals with favored features to produce offspring that exhibit the feature and that can be used for future matings. The desirable *phenotypic traits* are passed from parents to offspring via their *genes*

The term 'artificial selection' was first coined by *Charles Darwin* in his book *On the Origin of Species* to describe how humans had mirrored the process of natural selection through selective breeding.

Plants and Livestock

Almost all of the food consumed by modern humans has been selectively bred over thousands of years. Around 10,000 years ago when humans began living in permanent or semi-permanent settlements, they started to cultivate their own crops and herd flocks of livestock for the first time. Humans originally selected fruits and vegetables for qualities such as large size and sweetness unconsciously; the seeds of plants with the desirable qualities would have been given the chance to *germinate* through human consumption and cultivated within their latrines (toilets). Over time, other favorable qualities, such as oil content, seedlessness and fleshy texture, were all altered, leaving many human-cultivated fruit and vegetables unrecognizable against their wild counterparts. The same process occurred with domesticated animals such as sheep (bred for thicker wool), chickens (considerably larger than their wild ancestors), and cattle (bred for more muscle mass or increased milk yield).

One of the oldest and most widely documented examples of selective breeding for food is the selection of tall growing (for easier harvesting), disease resistant wheat, which yields large amounts of grain. Historically, smaller crops were removed from fields, allowing bees and other pollinators to pollinate only the crops with the most human-favored characteristics. Today, the breeding of wheat is a more scientific process; individuals with specific genes are identified and bred to create plants that have improved nutritional content and more intense flavors, and require less fertilizer or pesticide applications.

Selective breeding of domesticated animals has also resulted in the generation of diverse breeds of offspring:

- Examples of selective breeding of domesticated animals can be seen in horse and cows.

Example 1: Horse Breeding

Horses have been selectively bred across many generations to produce variation according to a targeted function:

- Race horses have been bred for speed and hence are typically leaner, lighter, taller and quicker.

- Draft horses have been bred for power and endurance and hence are sturdier and stockier.

Example 2: Cow Breeding

Cows have been selectively bred across many generations to produce offspring with improved milk production.

Farmers have also targeted the breeding a cows with a mutation resulting in increased muscle mass:

- The resulting stock of cattle (termed Belgian Blue) have excessive bulk and produce more edible lean meat.

Normal Cow

Belgian Blue

Inbreeding

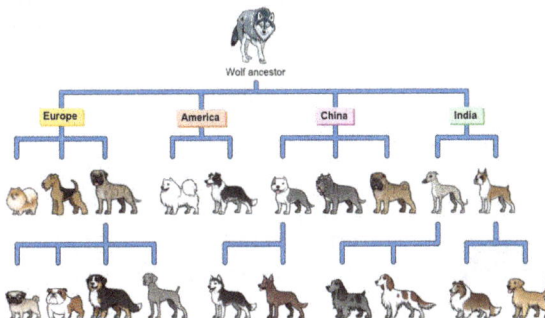

Various mating schemes of animals are classified under two broad categories — inbreeding and outbreeding. Classification depends on the closeness of the biological relationship between mates. Within each category, a wide variation in intensity of this relationship exists. A very fine line separates the two categories. Mating closely related animals (for example, parent and offspring, full brother and sister or half brother and sister) is inbreeding. With less closely related animals (first cousins, second cousins), people disagree about where to draw the line between inbreeding and outbreeding.

Technically, inbreeding is defined as the mating of animals more closely related than the average relationship within the breed or population concerned. Matings between animals less closely related than this, then, would constitute outbreeding. These two systems of mating, with varying intensities in each, are described in Table. Matings indicated within the inbreeding category are self-explanatory; those within the outbreeding category are defined in the glossary.

Degrees of inbreeding and outbreeding arranged according to biological relationship between indicated mates. (In reading from top to bottom, biological relationship between mates steadily decreases.)

Inbreeding or outbreeding	Coefficient of relationship between mates	Biological relationship between mates
Inbreeding	50 percent	Parent × offspring; full sibs
Inbreeding	25 percent	Half-sibs; double first cousins; aunt × nephew; uncle × niece
Inbreeding	12-1/2 percent	First cousins
Inbreeding	6-1/4 percent	Second cousins
Inbreeding	?	Linebreeding[1]
Inbreeding/ outbreeding	0 percent	Random mating within breed or population[2]
Outbreeding	0 percent	Outcrossing
Outbreeding	0 percent	Breed crossing
Outbreeding	0 percent	Species crossing
Outbreeding	0 percent	Genus crossing

[1]In a linebreeding program, the coefficient of relationship between mates is usually low; however, it can be quite variable.

[2]Random mating within a breed or population means that mates are chosen by chance. It should be understood that under this circumstance it is possible that either inbreeding or outbreeding could occur.

Biological Relationships between Animals

Individuals are considered to be biologically related when they have one or more common ancestors. For practical purposes, if two individuals have no common ancestor within the last five or six generations, they are considered unrelated.

Biological relationship is important in animal breeding because the closer the relationship, the higher the percentage of like genes the two individuals carry. Closeness of relationship is determined by three factors:

- How far back in the two animals' pedigrees the common ancestor appears

- How many common ancestors they have

- How frequently the common ancestors appear. It is also influenced by any inbreeding of the common ancestor or ancestors

Measurement of Degree of Biological Relationship

The coefficient of relationship is a single numerical value that considers all the above-mentioned factors. It is a measure of the degree to which the genotypes (genetic constitutions) of the two animals are similar. It is estimated by the expression:

$$R_{BC} = \text{sigma}[(1/2)^{n+n'}(1 + F_A)] \div \text{Square Root of } (1 + F_B)(1 + F_C)$$

Equation 1 where:

R_{BC} = the coefficient of relationship between animals B and C which we want to measure.

sigma = the Greek symbol meaning "add."

$(1/2)$ = the fraction of an individual's genetic material that is transmitted to its progeny. It is used in the calculation of the coefficient of relationshop because it represents the probability that, in any one generation, an identical gene from a given pair of genes is transitted to each of two particular progeny. It is also the probability that an unlike gene from a given pair of genes is transmitted to the two progeny.

n = the number of gnerations between animal B and the common ancestor.

n' = the number of generations between animal C and the common ancestor.

F_A, F_B, F_C = inbreeding coefficients of the common ancestor and of animals B and C, respectively.

If none of the animals is inbred, the coefficient of relationship is estimated as:

$$R_{BC} = \text{sigma} [(1/2)^{n + n'}]$$

Equation 2

The use of this expression can be demonstrated with the full-sib sample pedigree and arrow diagram. In this example, assume neither sire nor dam is inbred. The arrow diagram on the right shows paths of gene flow from each of the common ancestors (D and E) to the animals whose coefficient of relationship we are measuring (B and C).

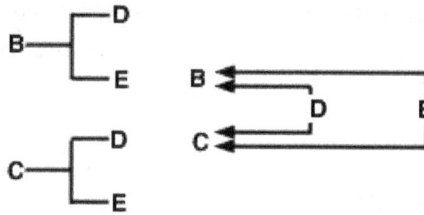

The problem now is to trace all possible paths from animal B to animal C which pass through a common ancestor. In this case, there are two such paths.

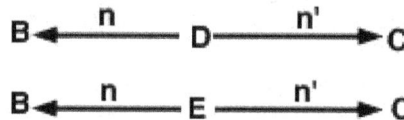

Equation 3

Since we have assumed no inbreeding in this example, the coefficient of relationship between animals B and C is estimated as:

$$RBC = \text{sigma}[(1/2)n + n'] = (1/2)1 + 1 + (1/2)1 + 1 = (1/2)2 + (1/2)2 = 0.50$$

Usefulness of Coefficient of Relationship Information

A livestock producer would find coefficient of relationship information valuable in a number of situations. He may, for example, want to sell an animal related to one that previously sold for a high price. The higher the coefficient of relationship between the two, the better its use as a sales point. Or, he may want to purchase one of two related bulls and one may cost more than he wishes to pay. If the coefficient of relationship between the two bulls is high, he might be as well off with the lower priced bull as he would with the more expensive one.

A practical use of the coefficient of relationship is estimating the performance value of an untested animal. To estimate the value, we must know the performance value of a related animal, the coefficient of relationship between the tested and untested animals, and the average performance value of the breed, herd, or group to which the tested and untested animals belong.

As an example, consider a herd with a feedlot average daily gain (ADG) of 2.25 pounds per day. Assume further that a sire from this herd had a 3.50 pounds per day feedlot ADG, while a younger half brother has not, as yet, been evaluated for feedlot ADG. Assuming no inbreeding, the coefficient of relationship between half brothers is 0.25. The best estimate of the untested animal's feedlot ADG is that it will deviate from the herd average 25 percent as far as does the performance value of the tested half brother. Using these figures, the most probable feedlot ADG value of the untested animal is 2.25 + (0.25) × (3.50 - 2.25) or 2.56 pounds per day.

Measurement of the Degree of Inbreeding

When we calculate an inbreeding coefficient, we are attempting to measure the probable percentage reduction in the frequency of pairing of dissimilar genes (reduction in heterozygosity). This reduction is relative to a base population. The base population usually is the breed concerned at a date to which the pedigrees are traced. Animals in this base population are assumed to be non-inbred. This does not mean these base population animals had dissimilar genes in each pair. There is no way for us to know how many of their gene pairs consisted of similar or dissimilar genes. The inbreeding coefficient that is calculated is simply relative to that base and reflects the probable percentage reduction in however many dissimilar gene pairs the average base population animals had.

The general expression for determining the inbreeding coefficient is:

$$F_X = \text{sigma}[(1/2)^{n+n'+1}(1 + F_A)]$$

Equation 4 where:

F_X = the inbreeding coefficient of animal X.

sigma = the Greek symbol meaning "add".

$(1/2)$ = the fraction of an individual's genetic material that is transmitted to its progeny. It is used in the calculation of the coefficient of relationshop because it represents the probability that, in any *one* generation, an identical gene from a given pair of genes is transitted to each of two particular progeny. It is also the probability that an *unlike* gene from a given pair of genes is transmitted to the two progeny.

n = the number of generations between animal B and the common ancestor.

n' = the number of generations between animal C and the common ancestor.

+1 = is added to n and n' to account for the additional generation between animal × and its parents.

F_A = the inbreeding coefficient of the common ancestor.

If neither parent is inbred, but if they are related, the inbreeding coefficient of their progeny is half their coefficient of relationship: $1/2\ R_{BC}$ This can be demonstrated using a full-sib mating to make comparison easy with the full-sib coefficient of relationship calculated previously. In this case, the pedigree for animal X and the arrow diagram will be as follows:

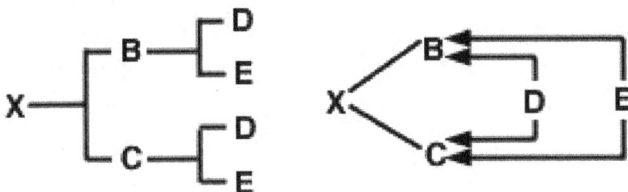

The problem now is to trace all possible paths from the sire (B) to the dam (C) through each common ancestor. As with the coefficient of relationship problem, there are two such paths:

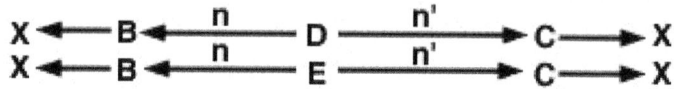

$$X \longleftarrow B \overset{n}{\underset{n}{\longleftarrow}} D \overset{n'}{\underset{n'}{\longrightarrow}} C \longrightarrow X$$
$$X \longleftarrow B \longleftarrow E \longrightarrow C \longrightarrow X$$

Since we have assumed neither parent is inbred, the inbreeding coefficient of animal × is estimated as:

$$F_X = \text{sigma}[(1/2)^{n+n'+1}] = (1/2)^{1+1+1} + (1/2)^{1+1+1} = (1/2)^3 + (1/2)^3 = 0.25$$

Equation 5

This is one-half the coefficient of relationship between full-sibs when there is no inbreeding.

Genetic Consequences of Inbreeding

The basic genetic consequence of inbreeding is to promote what is technically known as homozygosity. This means there is an increase in the frequency of pairing of similar genes. Accompanying this increase, there must be a decrease in the frequency of pairing of dissimilar genes. This is called a decrease in heterozygosity. These simultaneous events are the underlying reasons for the general effects on performance we observe with inbreeding.

Reasons for Inbreeding

Development of highly productive inbred lines of domestic livestock is possible. To date, however, such attempts have met with little apparent success. Although occasional high performance animals are produced, inbreeding generally results in an overall reduction in performance. This reduction is manifested in many ways. The most obvious effects of inbreeding are poorer reproductive efficiency including higher mortality rates, lower growth rates and a higher frequency of hereditary abnormalities. This has been shown by numerous studies with cattle, horses, sheep, swine and laboratory animals.

The extent of this decrease in performance, in general, is in proportion to the degree of inbreeding. The greater the degree of inbreeding, the greater the reduction in performance. The actual performance reduction is not the same in all species or in all traits. Some characteristics (like meat quality) are hardly influenced by inbreeding; others (like reproductive efficiency) are greatly influenced by inbreeding. We cannot, then, make a generalized statement about the amount of reduction in "performance" that would result from a specific amount of inbreeding and expect it to be applicable in a broad variety of situations.

It is possible, however, to predict the extent of the effect of inbreeding on specific traits. Such predictions are based on results actually obtained under experimental conditions in which various levels of inbreeding had been attained. In research with swine conducted at the Midwest Regional Swine Breeding Laboratory, Dickerson and others (1954) point out that for each 10 percent increase in inbreeding (of the pigs in the litter), there is a decrease of 0.20, 0.35, 0.38, and 0.44 pigs per litter at birth, 21 days, 56 days, and 154 days, respectively. We can use such figures to provide estimates of expected decreases in litter size (at comparable litter ages) in other herds of swine.

Most inbreeding studies suggest each successive unit increase in inbreeding results in a proportional decrease in performance. Estimates of average increases in percentage inbreeding within a closed herd can be made with the expression:

Avg. increase in inbreeding = (Number males + Number females) ÷ (8)(Number males)(Number females)

In a closed herd of cattle in which 100 females and four males were used in each generation, for example, the average per generation increase in inbreeding would be (4 + 100) ÷ (8)(4)(100) = 0.0325. On a per year basis, assuming a generation interval of five years, this would amount to an average yearly increase in inbreeding of 0.0065 or 0.65 percent.

Despite the generally poor results obtained with inbreeding, it is a very useful tool in animal breeding. Inbreeding is essential to the development of prepotent animals — animals that uniformly "stamp" their characteristics on their progeny. Because inbreeding causes an increase in the proportion of like genes (good or bad, recessive or dominant), the inbred animal's reproductive cells will be more uniform in their genetic makeup. When this uniformity involves a relatively large number of dominant genes, the progeny of that individual will uniformly display the dominant characteristics of that parent.

Inbreeding may also be used to uncover genes that produce abnormalities or death — genes that, in outbred herds, are generally present in low frequencies. These harmful genes are almost always recessive in their genetic nature and their effects are hidden or masked by their dominant counterparts (alleles). Except for sex-linked traits, recessive genes are not expressed if carried singly. For their effects to be manifested, they must be present in duplicate. The likelihood they will be present in duplicate increases with inbreeding, because inbreeding increases the proportion of like genes (both good and bad) in the inbred population. With the effects of these genes uncovered, the breeder can eliminate them from his herd. He would cull progeny that showed the undesirable effect of these recessive genes and would also cull the parents that are carriers of the undesirable genes. In addition, two-thirds of the "normal" progeny of these carrier parents are themselves expected to be carriers of these same undesirable genes. In the absence of breeding tests to sort out the carriers from the non-carriers, it would also be necessary to cull all "normal" progeny of the carrier parents.

Breeders can use an inbreeding test to identify carriers of harmful autosomal recessive genes (like those responsible for snorter dwarfism in cattle, hyperostosis in swine, or cryptorchidism in sheep). An inbreeding test checks for only recessive genes that the tested animal (usually the male) carries. The following example and the numbers given are pertinent to cattle, where one offspring per gestation is usual. If we want to be sure at the 0.01 level of probability that a bull is free of harmful autosomal recessive genes, we would have to mate him to at least 35 of his daughters. He is mated to 35+ of his daughters because only half of them are expected to carry any harmful recessive gene their sire is carrying. We need the production of 35 normal calves *without a single abnormal calf* to show the bull free (at the 0.01 probability level) of any harmful autosomal recessive gene.

Tests in sheep and swine would require matings to fewer daughters. Actual numbers would depend on the average number of progeny produced per gestation in each species. Using the 0.01 probability level and assuming the average litter size is 1.5 in sheep and 8.0 in swine, the number of sire-daughter matings needed would be about 24+ for sheep and 5+ for swine.

Another important use of inbreeding is in the development of distinct families or inbred lines. Beginning with an initially diverse genetic population, inbreeding results in the formation of various lines, each differing genetically from the other. Continued inbreeding within these lines tends to change the frequency of some of the genes found in the initial population. For example, if a particular gene is present in only 1 percent of the animals in the initial population, inbreeding and the development of distinct lines could result in this gene being present in all or nearly all animals in some lines and in none or only a few of the animals in other lines. Inbred lines are used in a number of ways but are probably most notably used in the development of hybrid chickens or hybrid seed corn.

A generally mild form of inbreeding (linebreeding) is being used successfully by some seed stock and commercial producers. Its objective is to maintain a high degree of relationship between the animals in the herd and some outstanding ancestor or ancestors. With inbreeding in general, there is no attempt to increase the relationship between the offspring and any particular ancestor. In a linebreeding program there is a deliberate attempt to maintain or increase the relationship between the offspring and a specific admired ancestor (or ancestors). This feature distinguishes linebreeding as a special form of inbreeding.

The inbreeding coefficient of the offspring produced in a linebreeding program is generally low, but is dependent on the kind of breeding program followed. Figure illustrates two linebreeding programs. The breeding program outlined in part A shows a direct line of relationship between offspring "x" and desired ancestor "5." Only a mild level of inbreeding in animal "x" ($F_x = 0.03125$) is reached. The coefficient of relationship obtained between animal "x" and animal "5" is 0.2462. Part B is a linebreeding program based on continuous sire-daughter matings which, after only two generations, produces a level of inbreeding in animal "x" of 0.375 and a coefficient of relationship between animals "x" and "S" of 0.78.

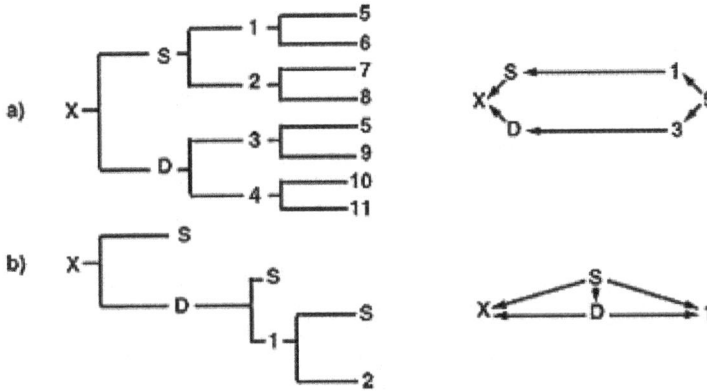

Pedigrees and arrow diagrams for two line breeding programs

The sire-daughter program in part B effectively concentrates genes from animal "S," but because of the rapid increase in inbreeding of the progeny produced in this program, the breeder runs the risk of greatly reduced performance and a high probability of genetic defects. Linebreeding programs are best used in purebred and high performing herds when a truly genetically superior individual in that herd has been identified and evaluated by progeny testing. Concentrating that individual's genes would then be best accomplished by mating him to unrelated females to reduce the risk of harmful effects associated with such intense inbreeding.

Outbreeding

Out-crossing or out-breeding means that the crossing between different breeds. This is the practice of introducing unrelated genetic material into a breeding line. It increases genetic diversity, thus reducing the probability of an individual being subject to disease or genetic abnormalities.

Outcrossing is now the norm of most purposeful animal breeding, contrary to what is commonly believed. The outcrossing breeder intends to remove the traits by using "new blood". With dominant traits, one can still see the expression of the traits and can remove those traits whether one outcrosses, line breeds or inbreds. With recessive traits, outcrossing allows for the recessive traits to migrate across a population. The outcrossing breeder then may have individuals that have many deleterious genes that may be expressed by subsequent inbreeding. There is now a gamut of deleterious genes within each individual in many dog breeds.

Increasing the variation of genes or alleles within the gene pool may protect against extinction by stressors from the environment. For example, in this context, a recent veterinary medicine study tried to determine the genetic diversity within cat breeds.

Outcrossing is believed to be the "norm" in the wild. Outcrossing in plants is usually enforced by self-incompatibility.

Breeders inbreed within their genetic pool, attempting to maintain desirable traits and to cull those traits that are undesirable. When undesirable traits begin to appear, mates are selected to determine if a trait is recessive or dominant. Removal of the trait is accomplished by breeding two individuals known not to carry it.

Gregor Mendel used outcrossing in his experiments with flowers. He then used the resulting offspring to chart inheritance patterns, using the crossing of siblings, and backcrossing to parents to determine how inheritance functioned.

Charles Darwin, in his book The Effects of Cross and Self-Fertilization in the Vegetable Kingdom, came to clear and definite conclusions concerning the adaptive benefit of outcrossing. For example, he stated (on page 462) that "the offspring from the union of two distinct individuals, especially if their progenitors have been subjected to very different conditions, have an immense advantage in height, weight, constitutional vigor and fertility over the self-fertilizing offspring from either one of the same parents". He thought that this observation was amply sufficient to account for outcrossing sexual reproduction. The disadvantages of self-fertilized offspring (inbreeding depression) are now thought to be largely due to the homozygous expression of deleterious recessive mutations; and the fitness advantages of outcrossed offspring are thought to be largely due to the heterozygous masking of such deleterious mutations.

Outbreeding Depression

In biology, outbreeding depression is when progeny resulting from crosses between genetically distant individuals (outcrossing) exhibit lower fitness in the parental environment than either of their parents or than progeny from crosses between individuals that are more closely related. The concept is opposed to inbreeding depression, although the two effects can occur simultaneously. Outbreeding depression manifests most significantly in two ways:

- Intermediate genotypes are not adapted to either parental habitat. For example, selection in one population might favor a large body size, whereas in another population small body size might be more advantageous, while individuals with intermediate body sizes are comparatively disadvantaged in both populations. As another example, in the Tatra Mountains, the introduction of ibex from the Middle East resulted in hybrids which produced calves at the coldest time of the year.

- Breakdown of biochemical or physiological compatibility. Within isolated breeding populations, alleles are selected in the context of the local genetic background. Because the same alleles may have rather different effects in different genetic backgrounds, there is the potential evolution of different locally adapted gene complexes. Outcrossing between individuals with differently adapted gene complexes can result in disruption of this selective advantage, resulting in a loss of fitness.

Mechanism and Impact

The different mechanisms of outbreeding depression can operate at the same time. However, determining which mechanism is more important in a particular population is very difficult. Generally the first mechanism will be more prevalent in the first generation (F1) after the initial outcrossing when most individuals are made up of the intermediate phenotype. An extreme case of this type of outbreeding depression is the sterility and other fitness-reducing effects often seen in interspecific hybrids (such as mules), which involves not only different alleles of the same gene but even different orthologous genes.

The second mechanism may not appear until two or more generations later (F2 or greater), when recombination has undermined vitality positive epistasis. Hybrid vigor in the first generation can, in some circumstances, be strong enough to mask the effects of outbreeding depression. An example of this is that plant breeders will make F1 hybrids from purebred strains, which will improve the uniformity and vigor of the offspring, however the F1 generation are not used for further breeding because of unpredictable phenotypes in their offspring. Unless there is strong selective pressure, outbreeding depression can increase in further generations as co-adapted gene complexes are broken apart without the forging of new co-adapted gene complexes to take their place.

If the outcrossing is limited and populations are large enough, selective pressure acting on each generation can restore fitness. Unless the F1 hybrid generation is sterile or very low fitness, selection will act in each generation using the increased diversity to adapt to the environment. This can lead to recovery in fitness to baseline, and sometimes even greater fitness than original parental types in that environment. However, as the hybrid population will likely to go through a decline in fitness for a few generations, they will need to persist long enough to allow selection to act before they can rebound.

Cross Breeding

Crossbreeding is the deliberate mating of animals from different breeds or strains designed to take advantage of Heterosis (Hybrid Vigour) for characteristics like production, fertility and longevity.

Heterosis (Hybrid Vigour)

Heterosis/Hybrid Vigour is the tendency of a crossbred animal to have qualities superior to that of either parent but not more than the dominant breed.

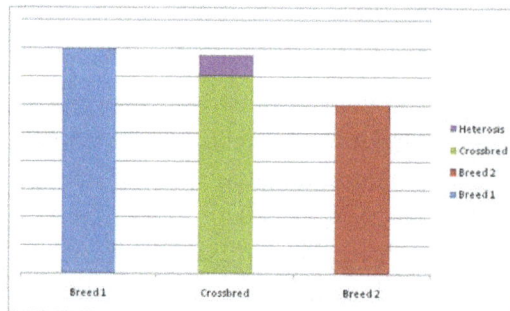

The above example of heterosis can be applied to all health or production traits if there is suffcient genetic difference between the two chosen breeds. For the above example if Breed 1 was crossed with Breed 2 the green area of the crossbred would be the logical average animal produced by the two breeds. The purple area on top of this is the added benefit or heterosis effect. The heterosis effect is variable between breeds and also the traits affected.

Trial work has shown the main benefit of heterosis is seen in fitness and health traits. 10% to 18% heterosis has been recorded in trials carried out in California and New Zealand in a wide variety of crossbred cows including Swedish/Norwegian Reds (SRB), Brown Swiss, Jerseys, and Montbelliarde. Milk production has been recorded as improving up to 6% but it is important to remember this is not the only measure of profitability.

To gain and maintain the effects of heterosis, breed and bull selection is paramount. To maximise the benefits farmers need to plan the direction they want their business to take in the long term to ensure cross breeding is the right path for them. Farmers initially need to look at the traits they wish to improve within their herd and then select bulls using all avaialble bull proofs including PLI and PIN.

2 Way Cross	67%
3 Way Cross	85%
4 Way Cross	94%

Although working a four way cross system provides maximum gain from heterosis, it is widly recomended a three way cross is more sustainable. Using a four way cross is:

- Difficult to implement
- Choosing four suitable breeds can be a problem
- Breed qualities are watered down with each cross

Heterosis should not be used as the main reason to crossbreed but as an added bonus. It is important to remember that crossbreeding is about herd improvement not breed improvement.

References

- Lewontin & Birch, R.C. & L.C. (February 3, 1966). "Hybridization as a source of variation for adaptation to new environments". Evolution. 20 (3): 315–336. doi:10.2307/2406633. JSTOR 2406633. PMID 28562982

- Frankham, Ballou, & Briscoe, R., J.D. & D.A. (2002). Introduction to Conservation Genetics. Cambridge. p. 382. ISBN 0521702712

- Bernstein H, Hopf FA, Michod RE (1987). "The molecular basis of the evolution of sex". Adv. Genet. 24: 323–70. doi:10.1016/s0065-2660(08)60012-7. PMID 3324702

- Jump up^ Turcek, FJ (1951). "Effect of introductions on two game populations in Czechoslovakia". Journal of Wildlife Management. 15: 113–114

- Fenster, Charles (2000). "Inbreeding and Outbreeding Depression in Natural Populations of Chamaecrista fasciculata (Fabaceae)". Conservation Biology. 14 (5): 1406–1412. doi:10.1046/j.1523-1739.2000.99234.x

Chapter 3

Poultry Farming and Management

Poultry farming refers to the rearing of poultry, which include birds like turkeys, chicken, ducks and geese for their meat and eggs. In this chapter, the significant concepts of poultry feeding, housing and various techniques of poultry management, such as debeaking and dubbing are explained in detail.

Poultry Farming

Poultry farming means 'raising various types of domestic birds commercially for the purpose of meat, eggs and feather production'. The most common and widely raised poultry birds are chicken. About 5k million chickens are being raised every year as a source of food (both meat and eggs of chicken). The chickens which are raised for eggs are called layer chicken, and the chickens which are raised for their meat production are called broiler chickens.

The UK and USA consume more meat and eggs of chicken than other countries of the world. On an average the UK alone consumes more than 29 million chicken eggs everyday. However, in a word commercial poultry farming is very necessary to meet up the demand of animal nutrition (eggs and meat). Commercial poultry farming is also very profitable. And commercial poultry farming business is one of the traditional business ventures. Here we are describing more about the advantages of poultry farming business and the steps for running this business.

Benefits of Poultry Farming

Poultry farming business has numerous benefits. As a result many farmers prefer to invest in this business. People generally establish poultry farm for the purpose of producing eggs, meat and generating high revenue from these products. Billions of chickens are being raised throughout the world as a good source of food from their eggs and meat. However, here I am shortly describing the main benefits of poultry farming.

- The main benefit of poultry farming is, it doesn't require high capital for starting. You need just basic capital to start raising poultry. And most of the poultry birds are not costly enough to start raising.

- Poultry farming doesn't require a big space unless you are going to start commercially. You can easily raise some birds on your own backyard with one or numerous coops or cages. So, if you are interested in poultry farming, then you can easily do it on your own backyard with several birds.

- Commercial poultry farming business also ensure high return of investment within a very short period. Some poultry birds like broiler chickens take shorter duration of time to mature and generating profit.

- Poultry farm structures do not require high maintenance. You can minimize diseases and illness in poultry by following proper hygiene and care. Diseases are less in some poultry birds like quails, turkeys etc.

- In most cases, you don't need any license. Because almost all types of poultry birds are domestic. Although, if you need license from the relevant authority it is also easy for poultry.

- Poultry provides fresh and nutritious food and has a huge global demand. Global consumers of poultry products prefer them due to their nutrients and freshness. Poultry products are not much expensive and most of the people can afford those.

- Marketing poultry products is very easy. There is an established market for poultry products in almost all places of the world. So, you don't have to think about marketing your products. You can easily sell the products in your nearest local market.

- Poultry farming creates income and employment opportunities for the people. Unemployed educated youth can easily create a great income and employment opportunity for them by raising poultry commercially. Women and students can also do this business along with their daily activities.

- Almost all bank approve loans for this types of business venture. So, if you want to start this business commercially, then you can apply for loans to your local banks.

- There are many more benefits of poultry farming along with the above mentioned benefits. Start raising and you will gradually learn everything.

Various Methods of Poultry Farming

Worldwatch institute described that, "about 74% of total poultry meat and 68% of total poultry eggs produced from intensive poultry farming method. Free range farming is the other alternative method of intensive poultry farming. Free range farming method is used for large number of poultry birds with high stocking density. There are some basic differences between intensive and free range poultry farming. Intensive poultry farming method is a highly efficient system which saves, land, feed, labor and other resources and increases production. In this system the poultry farming environment is fully controlled by the farmer. So, it ensures continuous production throughout the year in any environment and seasons. Intensive poultry farming has some disadvantages too. Some people says intensive system creates health risks, abuse the animals and harmful for environment. On the other hand free range poultry farming method requires a large place fro raising the birds and the production is about the same as intensive method. However, in the case of both intensive and free range poultry farming method the producers must have to use nationally approved medications like antibiotics regularly to keep the poultry birds free from diseases.

Layer Poultry Farming

The poultry birds which are raised for egg production are called layer poultry. Commercial hen generally starts laying eggs at the age of 12-20 weeks. They start laying eggs regularly at their 25 weeks of age. After 70-72 weeks of age egg production of layer poultry get reduced. For commercial layer poultry farming, producers generally keep the hens for 12 months from their first laying period. And then sell them for slaughter purpose. Although chickens naturally survive for more than 6 years. For re-invigorating egg laying, the hens are force moulted in some countries. For commercial egg laying poultry farming systems, the environmental conditions are often automatically controlled by the producers. For a simple example, presence of light helps the bird for laying eggs earlier. So, the producers should provide more lightening period to increase the probability of beginning laying eggs. The egg-laying birds lays more eggs in warmer months than the cold months. So, keeping the temperature of the room moderate will be very helpful for better egg production. Some commercial egg laying chicken breeds can produce more than 300 eggs a year. Layer poultry are raised in various methods. The common and most popular layer poultry farming systems are described shortly below.

- Free Range Farming: Free range poultry farming means providing freely roaming facilities to the poultry birds for a certain period of a day. Although they are kept inside the house at night to keep them free from predators and adverse weather. In free range farming method the poultry birds generally roam freely throughout the whole day. Which means they spent half of their life outside the house. For free range poultry farming system select a suitable land which has the facilities of adequate drainage system, good ventilation, appropriate protection from prevailing winds, good protection from all types of predators and free from excessive cold, heat or dampness. Excessive cold, heat and damp is very harmful for poultry birds and reduce their productivity. This system also requires less feed than cage and barn systems. The poultry manure from free range farming used as fertilizer for crops directly. Although free range farming method is very suitable for poultry birds but it has some difficulties too. In this system the poultry birds can be victim of predators easily and may caught by various type of diseases.

- Organic Method: Organic layer poultry rearing system is also one type of free range farming system. But the main differences between the two systems are, in free range farming method a large numbers of poultry birds are raised together but in organic method a certain species of poultry bird are raised in small group with low stocking density. Organic laying system has some restrictions in the routine use of synthetic yolk colourants, water, feed, medications, other feed additives and obviously a smaller group size with low stocking density. In organic laying system the producer should keep highest 1000 poultry birds per hector and maximum 2000 birds in each house.

- Yarding Method: Yarding poultry farming method is such a method in which cows and chickens are raised together. The farmer make a fence in his yard and keep all the poultry birds and cattle there together. The birds and cattle has the freedom of movement inside the fence. It is a very popular system used by small farmer.

- Battery Cage Method: Battery cage layer poultry rearing method is one of the very common methods used in many countries. In this system usually small sized metal cages are used. Every cages can accommodate about 3 to 8 hens. The walls of the cages are generally made of mesh or solid metal and the floor is made of sloped wire mesh which allow the faeces to drop down. When the hens lays eggs, then all the eggs gather in the egg collecting conveyor belt of the cage. In this system food is provided in front of the hens by a long bisected metal or plastic pipe and water served to them by using overhead nipple systems. The cages are arranged in long rows in one above another system. There may have several floors in a single shade which can keep many even thousands of hens together. For reducing feather and vent pecking, the light intensity is generally kept lower than 10 lux. The battery cage method has some benefits. The main benefits of battery cage are listed below.

1.   It is very easy to care for the birds.

2.   Very easy to collect eggs.

3.   Cleaner eggs.

4.   Requires less feed to produce eggs.

5.   Thousands of hens may be housed in a specific floor space of the house.

6.   The birds suffers less by internal parasites.

7.   Labor cost is very low.

Besides those benefits battery cage system has some difficulties too. By rearing large number of hens in a small place the air inside the house may contain high ratio of CO_2. The hens can't get sufficient space to walk, flap their wings, stand or perch. For this reason they may suffer by frustration and boredom and their behaviors may change which affect their production. Battery cage system is banned in some countries because it is considerate as against the animal welfare.

Furnished Cage Method: Furnished cage method is a developed version of battery cage system. In this system the hens get more spaces and facilities than battery cage system. A furnished cage for hens should contain sufficient space for walk, perch, flap their wings, nest, special feed and water pot etc.

Broiler Poultry Farming

The poultry birds which are raised for commercial meat production are called broiler poultry. By using modern farming methods broiler chickens become suitable for consumption within their 5 to 6 weeks of age. However, see the common raising systems which are mostly used for commercial broiler poultry farming.

- Indoor Raising Methods: In this method broilers are kept inside a house. Rice hulls, wood shavings, peanut shells etc. are used as litter in the floor of the house. In this system the broilers are kept in a large and open house (known as growout houses) and they become suitable for consumption within their 5 to 6 weeks of age. This types of poultry houses are well equipped with mechanical systems for delivering the feed and water to the poultry birds. Well ventilation system, coolers and heaters are must. It is very important to keep the house always dry and clean. Generally a house of 400 feet long and 40 feet wide can accommodate about 20,000 birds. One-half square feet space is required per bird.

- Free-Range Methods: In free-range broiler farming methods the broilers are kept like the free-range layers. The broiler breeds which grows slowly(takes more than 8 weeks for reaching slaughter weight) are suitable for raising in this systems. The main facilities of free-range farming systems is that, it allows the birds scratching, foraging, pecking and outdoor exercise.

- Organic Farming Methods: Organic farming method is almost the same as free-range farming systems. But the main difference is that, in organic farming methods the birds are not allowed for randomly using of in-feed or in-water medications, other food additives and synthetic amino acids. This system is very suitable for the poultry breeds which reach slaughter weight slowly (around 12 weeks).

Some Popular Poultry Breeds for Farming

There are some popular and mostly raised poultry birds. Among them chickens, turkeys, quails etc. are mostly raised poultry birds.

- Chicken

- Turkey

- Quail

- Ostrich

- Duck

- Pigeon

- Peacock

Poultry farming, raising of birds domestically or commercially, primarily for meat and eggs but also for feathers. Chickens, turkeys, ducks, and geese are of primary importance, while guinea fowl and squabs (young pigeons) are chiefly of local interest. This article treats the principles and practices of poultry farming.

Poultry Farm

Types of Poultry

Mass production of chicken meat and eggs began in the early 20th century, but by the middle of that century meat production had outstripped egg production as a specialized industry. The market for chicken meat has grown dramatically since then, with worldwide exports reaching nearly 12.5 million metric tons (about 13.8 million tons) by the early 21st century.

Egg Production
Egg-producing hens (*Gallus gallus*) in an industrial henhouse. Under ideal lighting and heating conditions, female chickens may produce one egg every 23–26 hours

The breeds of chickens are generally classified as American, Mediterranean, English, and Asiatic. While there are hundreds of breeds in existence, commercial facilities rely on only a select few that meet the rigorous demands of industrial production. The single-comb White Leghorn, a Mediterranean breed widely used throughout the global egg industry, is a prolific layer that quickly reaches sexual maturity. The Cornish Cross, a hybrid of Cornish and White Rock, is one of the most-common breeds for industrial meat production and is esteemed for its compact size and rapid, efficient growth.

Small farms and backyard flocks utilize a much wider variety of breeds and hybrids. Common American breeds include the Plymouth Rock, the Wyandotte, the Rhode Island Red, and the New Hampshire, all of which are dual-purpose breeds that are good for both eggs and meat. The Asiatic Brahma, thought to have originated in the United States from birds imported from China, is popular for both its meat and its large brown eggs.

Rhode Island Red rooster. Over the 7,400-year history of chicken (*Gallus gallus*) domestication, roosters (male chickens) have been used as fighting animals as well as for breeding and meat production

Turkey

After World War II, turkey production became highly specialized, with larger flocks predominating. Turkeys are raised in great numbers in Canada where their ancestors still live wild, as well as in some parts of Europe, the United States, Mexico, and Brazil. A hybrid white turkey dominates commercial production, while the Broad Breasted Bronze, the Broad Breasted White, the White Holland, and the Beltsville Small White are common breeds for smaller farms. In breeding flocks, one tom is required per 8 or 10 hens, though the modern hybrid turkey is too large for natural breeding and must be artificially inseminated.

Modern turkey breeding and farming practices have significantly reduced both the amount of feed and the time required to produce a pound of turkey meat. In 12–14 weeks a hen turkey eats about 16 kg (35 pounds) of feed and reaches 6–9 kg (14–20 pounds). Toms require some 36 kg (80 pounds) of feed to reach a market weight of 16–19 kg (35–42 pounds) in 16–19 weeks. Smaller turkey broilers are marketed from 12 to 15 weeks of age. Turkeys can be raised on open land with automatic waterers, self-feeders, range shelters, heavy fencing, and rotated pastures; however, they are often "grown out" commercially in rearing houses under environmentally controlled conditions.

Turkey
Midget White turkeys, a domestic breed of small stature

Ducks and Geese

Duck raising is practiced on a limited scale in nearly all countries, usually as a small-farm enterprise, though some commercial plants do exist. Ducks are easily transported, can be raised in close confinement, and convert some waste products and scattered grain (e.g., by gleaning rice fields) to nutritious and very desirable eggs and meat. Khaki Campbell and Indian Runner ducks are prolific layers, each averaging 300 eggs per year. The Pekin duck, one of the most popular breeds in the United States, is used for both egg and meat production. Although the white-fleshed Aylesbury was once the favoured meat duck in England, disease and market competition from the yellow-fleshed Pekin duck have led to its decline.

Pekin Duck
Domesticated Pekin ducks (Anas platyrhynchos domestica). Pekin ducks are prolific
egg layers and are one of the most popular breeds in the United States

Goose raising is often a minor farm enterprise, though some European countries have large-scale goose-production facilities. The two outstanding meat breeds are the Toulouse, predominantly gray in colour, and the Embden (or Emden), which is white. The birds are raised for meat and eggs as well as for their down feathers. Geese do not appear to have attracted the attention of geneticists on the same scale as the meat chicken and the turkey, and no change in the goose industry comparable to that in the others has occurred or seems to be in prospect. In some commercial plants, geese are fattened by a special process of force-feeding, resulting in a considerable enlargement of their livers, which are sold as the delicacy foie gras.

Domestic Goose
Flock of domestic geese (Anser anser domesticus). Domestic geese are often kept as poultry for their eggs, meat, and feathers

Guinea Fowl and Squabs

Guinea fowl are raised as a sideline on a few farms in many countries and are eaten as gourmet items. In Italy there is a fairly extensive industry. The birds are often raised in yards with open-fronted shelters, and a number of varieties and species are utilized throughout the world. Guinea fowl are marketed in England at 16–18 weeks of age and in the United States at about 10–12 weeks. The market weight is usually about 1–1.5 kg (2.5–3.5 pounds), but food conversion is poor.

Guinea Fowl
Helmeted guinea fowl (Numida meleagris)

Pigeons are raised not only as messengers and for sport but also for the meat of their squabs (nestlings). Squab production, carried on locally, is rare in most countries with established poultry industries, though the meat is often marketed as a gourmet item.

Poultry Feeding

Poultry convert feed into food products quickly, efficiently, and with relatively low environmental impact compared with other livestock. The high rate of productivity of poultry results in relatively high nutrient needs. Poultry require the presence of at least 38 dietary nutrients in appropriate concentrations and balance. The nutrient requirement figures published in Nutrient Requirements of Poultry (National Research Council, 1994) are the most recent available and should be viewed as minimal nutrient needs for poultry. They are derived from experimentally determined levels after an extensive review of the published data. Criteria used to determine the requirement for a given nutrient include growth, feed efficiency, egg production, prevention of deficiency symptoms, and quality of poultry product. These requirements assume the nutrients are in a highly bioavailable form, and they do not include a margin of safety. Consequently, adjustments should be made based on bioavailability of nutrients in various feedstuffs. A margin of safety should be added based on the length of time the diet will be stored before feeding, changes in rates of feed intake due to environmental temperature or dietary energy content, genetic strain, husbandry conditions (especially the level of sanitation), and the presence of stressors (such as diseases or mycotoxins).

Water

Water is an essential nutrient. Many factors influence water intake, including environmental temperature, relative humidity, salt and protein levels of the diet, birds' productivity (rate of growth or egg production), and the individual bird's ability to resorb water in the kidney. As a result, precise water requirements are highly variable. Water deprivation for ≥12 hr has an adverse effect on growth of young poultry and egg production of layers; water deprivation for ≥36 hr results in a marked increase in mortality of both young and mature poultry. Cool, clean water, uncontaminated by high levels of minerals or other potential toxic substances, must be available at all times.

Energy Requirements and Feed Intake

The energy requirements of poultry and the energy content of feedstuffs are expressed in kilocalories (1 kcal equals 4.1868 kilojoules). Two different measures of the bioavailable energy in feedstuffs are in use, metabolizable energy (AMEn) and the true metabolizable energy (TMEn). AMEn is the gross energy of the feed minus the gross energy of the excreta after a correction for the nitrogen retained in the body. Calculations of TMEn make an additional correction to account for endogenous losses of energy that are not directly attributable to the feedstuff and are usually a more useful measure. AMEn and TMEn are similar for many ingredients. However, the two values differ substantially for some ingredients such as feather meal, rice, wheat middlings, and corn distiller's grains with solubles.

Poultry can adjust their feed intake over a considerable range of feed energy levels to meet their daily energy needs. Energy needs and, consequently, feed intake also vary considerably with environmental temperature and amount of physical activity. A bird's daily need for amino acids, vitamins, and minerals are mostly independent of these factors. The nutrient requirement values in the following tables are based on typical rates of intake of birds in a thermoneutral environment consuming a diet that contains a specific energy content (eg, 3,200 kcal/kg for broilers). If a bird consumes a diet that has a higher energy content, it will decrease its feed intake; consequently, that diet must contain a proportionally higher amount of amino acids, vitamins, and minerals. Thus, nutrient density in the ration should be adjusted to provide appropriate nutrient intake based on requirements and the actual feed intake.

Because of the ability of poultry to adjust their feed intake to accommodate a wide range of diets with differing energy content, the energy values listed in the nutrient requirement tables in this section (Nutrient Requirements of Growing Pullets a through Linoleic Acid, Mineral, and Vitamin Requirements of Turkeys) should be regarded as guidelines rather than absolute requirements.

Nutrient Requirements of Growing Pullets [a]

Age (wk)	0–6	6–12	12–18	18 to 1st Egg
White-Egg Layers				
Body weight (g)[b]	**450**	**980**	**1,375**	**1,475**
Protein	18	16	15	17
Arginine	1.0	0.83	0.67	0.75
Lysine	0.85	0.60	0.45	0.52
Methionine	0.30	0.25	0.20	0.22
Methionine + cystine	0.62	0.52	0.42	0.47
Threonine	0.68	0.57	0.37	0.47

Age (wk)	0–6	6–12	12–18	18 to 1st Egg
Tryptophan	0.17	0.14	0.11	0.12
Calcium	0.90	0.80	0.80	2.00
Phosphorus, available	0.40	0.35	0.30	0.32
Brown-Egg Layers				
Body weight (g)[b]	**500**	**1,100**	**1,500**	**1,600**
Protein	17	15	14	16
Arginine	0.94	0.78	0.62	0.72
Lysine	0.80	0.56	0.42	0.49
Methionine	0.28	0.23	0.19	0.21
Methionine + cystine	0.59	0.49	0.39	0.44
Threonine	0.64	0.53	0.35	0.44
Tryptophan	0.16	0.13	0.10	0.11
Calcium	0.90	0.80	0.80	1.8
Phosphorus, available	0.40	0.35	0.30	0.35

[a] Requirements are listed as percentages of diet. Nutrient levels should be adjusted to meet specific strain requirements, level of feed intake, and body weight and skeletal development.

[b] Average body weight at end of each period.

Nutrient Requirements of Laying Hens at Different Feed Intakes [a]

Pounds (approx.)/100 birds/day	18	20	22	24	26
Grams of feed/bird/day	80	90	100	110	120
White-Egg Layers					
Protein	18.8	16.7	15.0	13.6	12.5
Arginine	0.88	0.78	0.70	0.64	0.58
Lysine	0.86	0.77	0.69	0.63	0.58
Methionine	0.38	0.33	0.30	0.27	0.25
Methionine + cystine	0.73	0.64	0.58	0.53	0.48
Threonine	0.59	0.52	0.47	0.43	0.39
Tryptophan	0.20	0.18	0.16	0.15	0.13
Calcium	4.12	3.67	3.30	3.00	2.75
Phosphorus, available	0.31	0.28	0.25	0.23	0.21
Brown-Egg Layers					
Protein	22.5	20.0	18.0	16.4	15.0
Arginine	1.06	0.94	0.85	0.77	0.71

Pounds (approx.)/100 birds/day	18	20	22	24	26
Grams of feed/bird/day	80	90	100	110	120
Lysine	1.05	0.93	0.84	0.76	0.70
Methionine	0.45	0.40	0.36	0.33	0.30
Methionine + cystine	0.89	0.79	0.71	0.65	0.59
Threonine	0.71	0.63	0.57	0.52	0.48
Tryptophan	0.24	0.21	0.19	0.17	0.16
Calcium	5.00	4.44	4.00	3.64	3.33
Phosphorus, available	0.38	0.33	0.30	0.27	0.25
[a] Requirements are listed as percentages of diet.					

Nutrient Requirements of Broilers [a]

Age[b]	0–3 wk	3–6 wk	6–8 wk
kcal AME_n/kg diet[c]	3,200	3,200	3,200
Crude protein[d]	23.00	20.00	18.00
Arginine	1.25	1.10	1.00
Glycine + serine	1.25	1.14	0.97
Histidine	0.35	0.32	0.27
Isoleucine	0.80	0.73	0.62
Leucine	1.20	1.09	0.93
Lysine[e]	1.10	1.00	0.85
Methionine	0.50	0.38	0.32
Methionine + cystine	0.90	0.72	0.60
Phenylalanine	0.72	0.65	0.56
Phenylalanine + tyrosine	1.34	1.22	1.04
Proline	0.60	0.55	0.46
Threonine	0.80	0.74	0.68
Tryptophan	0.20	0.18	0.16
Valine	0.90	0.82	0.70
[a] Requirements are listed as percentages of diet.			
[b] The 0- to 3-, 3- to 6-, and 6- to 8-wk intervals for nutrient requirements are based on chronology for which research data were available; however, these nutrient requirements are often implemented at younger age intervals or on a weight-of-feed consumed basis.			
[c] These are typical dietary energy concentrations. Different energy values may be appropriate depending on local ingredient prices and availability.			
[d] Broiler chickens do not have a requirement for crude protein per se. However, there should be sufficient crude protein to ensure an adequate nitrogen supply for synthesis of nonessential amino acids. Suggested requirements for crude protein are typical of those derived with corn-soybean meal diets, and levels can be reduced when synthetic amino acids are used.			
[e] Recent research has shown that higher levels of lysine are needed for maximal growth and efficiency of modern broilers.			

Protein and Amino Acid Requirements of Turkeys [a]

	---------------Age (wk)---------------							
Male:	0–4	4–8	8–12	12–16	16–20	20–24		
Female:	0–4	4–8	8–11	11–14	14–17	17–20	Holding	Breeding Hens
Energy base kcal ME/kg diet[b]	2,800	2,900	3,000	3,100	3,200	3,300	2,900	2,900
Protein	28.0	26	22	19	16.5	14	12	14
Arginine	1.6	1.4	1.1	0.9	0.75	0.6	0.5	0.6
Glycine + serine	1.0	0.9	0.8	0.7	0.6	0.5	0.4	0.5
Histidine	0.58	0.5	0.4	0.3	0.25	0.2	0.2	0.3
Isoleucine	1.1	1.0	0.8	0.6	0.5	0.45	0.4	0.5
Leucine	1.9	1.75	1.5	1.25	1.0	0.8	0.5	0.5
Lysine	1.6	1.5	1.3	1.0	0.8	0.65	0.5	0.6
Methionine	0.55	0.45	0.4	0.35	0.25	0.25	0.2	0.2
Methionine + cystine	1.05	0.95	0.8	0.65	0.55	0.45	0.4	0.4
Phenylalanine	1.0	0.9	0.8	0.7	0.6	0.5	0.4	0.55
Phenylalanine + tyrosine	1.8	1.6	1.2	1.0	0.9	0.9	0.8	1.0
Threonine	1.0	0.95	0.8	0.75	0.6	0.5	0.4	0.45
Tryptophan	0.26	0.24	0.2	0.18	0.15	0.13	0.1	0.13
Valine	1.2	1.1	0.9	0.8	0.7	0.6	0.5	0.58

[a] Requirements are listed as percentages of diet.

[b] These are typical ME concentrations for corn-soya diets. Different ME values may be appropriate if other ingredients predominate.

Adapted, with permission, from *Nutrient Requirements of Poultry*, 1994, National Academy of Sciences, National Academy Press, Washington, DC.

Nutrient Requirements of Pheasants [a]

Energy base	0–4 wk	4–8 wk	9–17 wk	Breeding
kcal ME/kg diet[b]	2,800	2,800	2,700	2,800
Protein (%)	28	24	18	15
Glycine + serine (%)	1.8	1.55	1	0.5
Lysine (%)	1.5	1.40	0.8	0.68
Methionine + cystine (%)	1.0	0.93	0.6	0.6
Linoleic acid (%)	1	1	1	1
Calcium (%)	1.0	0.85	0.53	2.5
Phosphorus, available (%)	0.55	0.5	0.45	0.40
Sodium (%)	0.15	0.15	0.15	0.15

Energy base	0–4 wk	4–8 wk	9–17 wk	Breeding
kcal ME/kg diet[b]	2,800	2,800	2,700	2,800
Chlorine (%)	0.11	0.11	0.11	0.11
Iodine (mg)	0.3	0.3	0.3	0.3
Riboflavin (mg)	3.4	3.4	3.0	4.0
Pantothenic acid (mg)	10	10	10	16
Niacin (mg)	70	70	40	30
Choline (mg)	1,430	1,300	1,000	1,000

[a] Requirements are listed as percentages or as mg/kg of diet. For values not listed, see requirements of turkeys (Protein and Amino Acid Requirements of Turkeys [a] and Linoleic Acid, Mineral, and Vitamin Requirements of Turkeys [a]) as a guide.

[b] These are typical dietary energy concentrations.

Adapted, with permission, from *Nutrient Requirements of Poultry*, 1994, National Academy of Sciences, National Academy Press, Washington, DC.

Nutrient Requirements of Bobwhite Quail [a]

Energy base kcal ME/kg diet[b]	Starting 2,800	Growing 2,800	Breeding 2,800
Protein (%)	26	20	24
Glycine + serine (%)	—	—	—
Lysine (%)	—	—	—
Methionine + cystine (%)	1.0	0.75	0.90
Linoleic acid (%)	1	1	1
Calcium (%)	0.65	0.65	2.4
Phosphorus, available (%)	0.45	0.30	0.7
Sodium (%)	0.15	0.15	0.15
Chlorine (%)	0.11	0.11	0.11
Iodine (mg)	0.3	0.3	0.3
Riboflavin (mg)	3.8	3.0	4.0
Pantothenic acid (mg)	12	9	15
Niacin (mg)	30	30	20
Choline (mg)	1,500	1,500	1,000

[a] Requirements are listed as percentages or as mg/kg of diet. For values not listed, see requirements of laying hens (Nutrient Requirements of Laying Hens at Different Feed Intakes [a]) and leghorn-type chickens (Linoleic Acid, Mineral, and Vitamin Requirements of Leghorn-type Chickens [a]) as a guide.

[b] These are typical dietary energy concentrations.

Adapted, with permission, from *Nutrient Requirements of Poultry*, 1994, National Academy of Sciences, National Academy Press, Washington, DC.

Nutrient Requirements of Pekin Ducks [a]

Energy base kcal ME/kg diet[b]	Starting (0–2 wk) 2,900	Growing (2–7 wk) 3,000	Breeding 2,900
Protein (%)	22	16	15
Arginine (%)	1.1	1.0	—
Lysine (%)	0.9	0.65	0.6
Methionine + cystine (%)	0.7	0.55	0.5
Calcium (%)	0.65	0.6	2.75
Phosphorus, available (%)	0.40	0.30	0.30
Sodium (%)	0.15	0.15	0.15
Chlorine (%)	0.12	0.12	0.12
Magnesium (mg)	500	500	500
Manganese (mg)	50	?	?
Zinc (mg)	60	?	?
Selenium (mg)	0.2	?	?
Vitamin A (IU)	2,500	2,500	4,000
Vitamin D (IU)	400	400	900
Vitamin K (mg)	0.5	0.5	0.5
Riboflavin (mg)	4	4	4
Pantothenic acid (mg)	11	11	11
Niacin (mg)	55	55	55
Pyridoxine (mg)	2.5	2.5	3.0

[a] Requirements are listed as percentages or as units or mg/kg of diet. For nutrients not listed, see nutrient requirements of broilers (Nutrient Requirements of Broilers [a]) as a guide.

[b] These are typical dietary energy concentrations.

Adapted, with permission, from *Nutrient Requirements of Poultry*, 1994, National Academy of Sciences, National Academy Press, Washington, DC.

Nutrient Requirements of Geese [a]

Energy base kcal ME/kg diet[b]	Starting (0–4 wk) 2,900	Growing (after 4 wk) 3,000	Breeding 2,900
Protein (%)	20	15	15
Lysine (%)	1.0	0.85	0.6
Methionine + cystine (%)	0.6	0.5	0.5
Calcium (%)	0.65	0.6	2.25
Phosphorus, available (%)	0.3	0.3	0.3
Vitamin A (IU)	1,500	1,500	4,000
Vitamin D (IU)	200	200	200
Riboflavin (mg)	3.8	2.5	4.0
Pantothenic acid (mg)	15	10	10

Energy base kcal ME/kg diet[b]	Starting (0–4 wk) 2,900	Growing (after 4 wk) 3,000	Breeding 2,900
Niacin (mg)	65	35	20
[a] Requirements are listed as percentages or as units or mg/kg of diet. For nutrients not listed, see requirements of broilers (Nutrient Requirements of Broilers [a]) as a guide.			
[b] These are typical dietary energy concentrations.			
Adapted, with permission, from *Nutrient Requirements of Poultry*, 1994, National Academy of Sciences, National Academy Press, Washington, DC.			

Linoleic Acid, Mineral, and Vitamin Requirements of Leghorn-type Chickens [a]

Age	0–6 wk	6–18 wk	18 wk to 1st egg	Layers	Breeders
Linoleic acid (%)	1.00	1.00	1.00	1.00	1.00
Potassium (%)	0.25	0.25	0.25	0.15	0.15
Sodium (%)	0.15	0.15	0.15	0.15	0.15
Chlorine (%)	0.15	0.15	0.15	0.13	0.13
Magnesium (mg)	600	500	400	500	500
Manganese (mg)	60	30	30	20	20
Zinc (mg)	40	35	35	35	45
Iron (mg)	80	60	60	45	60
Copper (mg)	5	4	4	?	?
Iodine (mg)	0.35	0.35	0.35	0.035	0.01
Selenium (mg)	0.15	0.1	0.1	0.06	0.06
Vitamin A (IU)	1,500	1,500	1,500	3,000	3,000
Vitamin D$_3$ (IU)	200	200	300	300	300
Vitamin E (IU)	10	5	5	5	10
Vitamin K (mg)	0.5	0.5	0.5	0.5	1.0
Riboflavin (mg)	3.6	1.8	2.2	2.5	3.6
Pantothenic acid (mg)	10	10	10	2	7
Niacin (mg)	27	10	10	10	10
Vitamin B$_{12}$ (mg)	0.009	0.003	0.004	0.004	0.08
Choline (mg)	1,300	900	500	1,050	1,050
Biotin (mg)	0.15	0.1	0.1	0.1	0.1
Folacin (mg)	0.55	0.25	0.25	0.25	0.35
Thiamine (mg)	1.0	1.0	0.8	0.7	0.7
Pyridoxine (mg)	3	3	3	2.5	4.5
[a] Requirements are listed as percentages or as units or mg/kg of diet. Assumes an average daily intake of 110 g of feed/hen/day.					
Adapted, with permission, from *Nutrient Requirements of Poultry*, 1994, National Academy of Sciences, National Academy Press, Washington, DC.					

Linoleic Acid, Mineral, and Vitamin Requirements of Turkeys [a]

	----------------Age (wk)---------------							
Male:	0–4	4–8	8–12	12–16	16–20	20–24		
Female:	0–4	4–8	8–11	11–14	14–17	17–20	Holding	Breeding Hens
Energy base kcal ME/ kg diet[b]	2,800	2,900	3,000	3,100	3,200	3,300	2,900	2,900
Linoleic acid (%)	1.0	1.0	0.8	0.8	0.8	0.8	0.8	1.1
Calcium (%)	1.2	1.0	0.85	0.75	0.65	0.55	0.5	2.25
Phosphorus, available (%)	0.6	0.5	0.42	0.38	0.32	0.28	0.25	0.35
Potassium (%)	0.7	0.6	0.5	0.5	0.4	0.4	0.4	0.6
Sodium (%)	0.17	0.15	0.12	0.12	0.12	0.12	0.12	0.12
Chlorine (%)	0.15	0.14	0.14	0.12	0.12	0.12	0.12	0.12
Magnesium (mg)	500	500	500	500	500	500	500	500
Manganese (mg)	60	60	60	60	60	60	60	60
Zinc (mg)	70	65	50	40	40	40	40	65
Iron (mg)	80	60	60	60	50	50	50	60
Copper (mg)	8	8	6	6	6	6	6	8
Iodine (mg)	0.4	0.4	0.4	0.4	0.4	0.4	0.4	0.4
Selenium (mg)	0.2	0.2	0.2	0.2	0.2	0.2	0.2	0.2
Vitamin A (IU)	5,000	5,000	5,000	5,000	5,000	5,000	5,000	5,000
Vitamin D[c] (IU)	1,100	1,100	1,100	1,100	1,100	1,100	1,100	1,100
Vitamin E (IU)	12	12	10	10	10	10	10	25
Vitamin K (mg)	1.75	1.5	1.0	0.75	0.75	0.5	0.5	1.0
Riboflavin (mg)	4.0	3.6	3.0	3.0	2.5	2.5	2.5	4.0
Pantothenic acid (mg)	10	9	9	9	9	9	9	16
Niacin (mg)	60	60	50	50	40	40	40	40
Vitamin B_{12} (mg)	0.003	0.003	0.003	0.003	0.003	0.003	0.003	0.003
Choline (mg)	1,600	1,400	1,100	1,100	950	800	800	1,000
Biotin (mg)	0.2	0.2	0.125	0.125	0.100	0.100	0.100	0.2
Folacin (mg)	1.0	1.0	0.8	0.8	0.7	0.7	0.7	1.0
Thiamine (mg)	2	2	2	2	2	2	2	2
Pyridoxine (mg)	4.5	4.5	3.5	3.5	3.0	3.0	3.0	4.0

[a] Requirements are listed as percentages or as units or mg/kg of diet.

[b] These are typical ME concentrations for corn-soya diets. Different ME values may be appropriate if other ingredients predominate.

[c] These concentrations of vitamin D are satisfactory when the dietary concentrations of calcium and available phosphorus conform with those in this table.

Adapted, with permission, from *Nutrient Requirements of Poultry*, 1994, National Academy of Sciences, National Academy Press, Washington, DC.

Appropriate body weight and fat deposition are important factors in rearing pullets for maximal egg production. Most strains of White Leghorn chickens have relatively low body weights and do not tend, under normal feeding, to become obese. Feed is normally provided for ad lib intake to this strain of pullets. For brown-egg strains of chickens, some degree of restriction is often practiced (~90% of ad lib feeding) to prevent precocial onset of lay. Broiler strains tend to become obese if fed ab lib; feed restriction is necessary for broiler pullets and broiler breeders. When feed restriction is practiced, the feed levels of amino acids, vitamins, and minerals must be proportionally increased to prevent deficiencies. Most large commercial breeders provide feed restriction and dietary nutrient guidelines specific for their strains.

Amino Acid Requirements

Poultry, like all animals, synthesize proteins that contain 20 L-amino acids. Birds are unable to synthesize 9 of these amino acids because of the lack of specific enzymes: arginine, isoleucine, leucine, lysine, methionine, phenylalanine, threonine, tryptophan, and valine. Histidine, glycine, and proline can be synthesized by birds, but the rate is usually insufficient to meet metabolic needs and a dietary source is required. These 12 amino acids are referred to as the essential amino acids. Tyrosine and cysteine can be synthesized from phenylalanine and methionine, respectively, and are referred to as conditionally essential because they must be in the diet if phenylalanine or methionine levels are inadequate. The diet must also supply sufficient amounts of nitrogen to allow the synthesis of nonessential amino acids. Essential amino acids are often added to the diet in purified form (eg, DL-methionine and L-lysine) to minimize the total protein level as well as the cost of the diet. This has the added advantage of minimizing nitrogen excretion.

Vitamins

Requirements for vitamins A, D, and E are expressed in IU. For chickens, 1 IU of vitamin A activity is equivalent to 0.3 mcg of pure retinol, 0.344 mcg of retinyl acetate, or 0.6 mcg of β-carotene. However, young chicks use β-carotene less efficiently.

One IU of vitamin D is equal to 0.025 mcg of cholecalciferol (vitamin D_3). Ergocalciferol (vitamin D_2) is used with an efficiency of <10% of vitamin D_3 in poultry.

One IU of vitamin E is equivalent to 1 mg of synthetic dl-α-tocopherol acetate. Vitamin E requirements vary with type and level of fat in the diet, the levels of selenium and trace minerals, and the presence or absence of other antioxidants. When diets high in long-chain highly polyunsaturated fatty acids are fed, vitamin E levels should be increased considerably.

Choline is required as an integral part of the body phospholipid, as a part of acetylcholine, and as a source of methyl groups. Growing chickens can also use betaine as a methylating agent. Betaine is widely distributed in practical feedstuffs and can spare the requirement for choline but cannot completely replace it in the diet.

All vitamins are subject to degradation over time, and this process is accelerated by moisture, oxygen, trace minerals, heat, and light. Stabilized vitamin preparations and generous margins of safety are often applied to account for these losses. This is especially true if diets are pelleted, extruded, or stored for long periods.

Minerals

Much of the phosphorus in feedstuffs of plant origin is complexed by phytate and is not absorbed efficiently by poultry. Consequently, it is critical that only the available phosphorus and not the total phosphorus levels be considered. Appropriate calcium nutrition depends on both the level of calcium and its ratio to that of available phosphorus. For growing poultry, this ratio should not deviate substantially from 2:1. The calcium requirement of laying hens is very high and increases with the rate of egg production and age of the hen.

Other Nutrients and Additives

The chick has requirements for 38 nutrients, together with an adequate level of metabolizable energy and water. Some additional nutrients may be necessary for growth and development under certain conditions. These include vitamin C, pyrroloquinoline quinone, and several heavy metals.

Non-nutrient antioxidants, such as ethoxyquin, are usually added to poultry diets to protect vitamins and unsaturated fatty acids from oxidation. Antibiotics at low levels (5–25 mg/kg of feed, depending on the antibiotic) and surfeit copper (150 ppm) are sometimes included to improve growth rate and feed efficiency. Enzymes that increase the bioavailability of dietary phosphorus, energy, and protein are often used in poultry diets when their costs are not prohibitive. In some cases, phytase enzymes are used to decrease the amount of phosphorus in the excreta to meet environmental regulations.

Poultry Feed Ingredients

Feed ingredients for poultry diets are selected for the nutrients they can provide, the absence of anti-nutritional or toxic factors, their palatability or effect on voluntary feed intake, and their cost. The key nutrients that need to be supplied by the dietary ingredients are amino acids contained in proteins, vitamins and minerals. All life functions also require energy, obtained from starches, lipids and proteins.

Feed ingredients are broadly classified into cereal grains, protein meals, fats and oils, minerals, feed additives, and miscellaneous raw materials, such as roots and tubers. These will be discussed in separate headings below. More information on measuring the nutrient composition of ingredients and the process of formulating poultry feeds is available in the section on feed formulation.

Cereal Grains

The term "cereal gains" here includes cereal grains, cereal by-products and distillers dry grains with solubles (DDGS). Cereal grains are used mainly to satisfy the energy requirement of poultry. The dominant feed grain is corn, although different grains are used in various countries and regions of the world. For instance, in the US, Brazil and most Asian countries corn is by far the most important energy source for all poultry feed, whereas wheat is the predominant supplier of dietary energy for poultry diets in Europe, Canada, Australia, New Zealand and the Russian Federation. Of course, in reality, a feed manufacturer will use any grain in a poultry diet if it is available at a reasonable price. For instance, in some parts of the US and China wheat is often used in place of corn if its price is below that of corn. In Australia, sorghum is a key grain during the summer season instead of wheat, while in the Scandinavian countries barley and rye are used when these grains are at the right price. Although the amounts and types of cereal grains included in poultry diets will depend largely on their current costs relative to their nutritive values, care must be taken to avoid making large changes to the cereal component of diets as sudden changes can cause digestive upsets that may reduce productivity and predispose the birds to disease.

Corn (maize)   Wheat   Sorghum

Table: ME value and key nutrient composition of cereal grains

Ingredient	Protein (%)	ME (kcal/kg)	Calcium (%)	Available P (%)	Lysine (%)
Wheat	13.0	3153	0.05	0.20	0.5
Corn	8.5	3300	0.05	0.20	0.3
Sorghum	9.0	3263	0.02	0.15	0.3
Barley	11.5	2795	0.10	0.20	0.4
Rye	12.5	2734	0.05	0.18	0.5
Triticale	15.4	3110	0.05	0.19	0.4
Oats	12.0	2756	0.10	0.20	0.4

The quality of cereal grains will also depend on seasonal and storage conditions. Poor growing or storage conditions can lead to grains with a lower than expected energy content or contamination with mycotoxins or toxin-producing organisms such as fungi and ergots. Genetic and environmental factors also affect not only the content of nutrients in grains but also the nutritive value, which takes into account the digestibility of nutrients contained in an ingredient in the target animal.

In addition to the cereals themselves, their by-products, such as wheat bran, rice bran and DDGS, are used widely in poultry feed. Cereal by-products are typically high in fibre, or non-starch polysaccharides (NSP), which are poorly utilised in poultry and are low in ME.

Protein Meals

Protein is provided from both vegetable and animal sources, such as oilseed meals, legumes and abattoir and fish processing by-products.

Vegetable Protein Sources

Vegetable protein sources usually come as meal or cake, the by-product of oilseed crops. The main oilseed crops include soybean, rapeseed/canola, sunflower, palm kernel, copra, linseed peanut and sesame seed. After the oil is extracted, the remaining residue is used as feed ingredient. Oilseed meals make up 20-30% of a poultry diet. Inclusion levels do vary among formulations for different species and for the same species in different regions.

Soybean   Canola seed   Lupins
                                                   (lupinus angustifolius)

The main vegetable protein sources used in Australian poultry diets are soybean and canola. Other sources like cottonseed, sunflower, peas and lupins may be included in poultry feed formulations if these are available at a reasonable price.

Many oilseeds and legumes contain anti-nutritive factors. Some of these anti-nutritive factors can be destroyed by heat and are used in heat-treated meals. New cultivars of some oilseeds and legumes have been developed that are naturally low in anti-nutritive factors (ANF), permitting higher levels of the unprocessed grains to be included in poultry diets without ill-effect. The typical energy values and nutrient composition of vegetable protein sources are shown in Table.

Table: ME values and Nutrient composition of vegetable protein sources

Ingredient	Protein (%)	ME (kcal/kg)	Calcium (%)	Available P (%)	Lysine (%)	Main Anti-nutrition-al factor
Soybean meal	48.0	2557	0.20	0.37	3.2	Trypsin inhibitor
Canola meal	37.5	2000	0.66	0.47	2.2	Glucosinolates
Cottonseed meal	41.0	2350	0.15	0.48	1.7	Gossypol
Sunflower meal	46.8	2205	0.30	0.50	1.6	High fibre
Peas	23.5	2550	0.10	0.20	1.6	Trypsin inhibitor
Lupins	34.5	3000	0.20	0.20	1.7	Toxic alkaloid

Animal Protein Sources

The main animal protein sources used in poultry diets are meat meal, meat and bone meal, fish meal, poultry by-product meal, blood meal and feather meal. Although the production of animal protein for human consumption has been under continual pressure and marred by much controversy, the world-wide and domestic consumption of animal protein continues to grow and much of the future supply of meat protein will come from poultry. With increased animal protein production there will be increased demand for feed and, in particular, a demand for ingredients high in protein and energy.

The animal industry evolved as a means of adding value (i.e. higher nutrient level and availability, flavour, variety, etc.) to ingredients that were of marginal food value for humans. These ingredients include grains that are of poor quality or damaged by harvest or storage conditions; as well as a means of recycling by-products of brewing, vegetable oil, meat, milk and egg production. Approximately 50% of the live market weight of ruminants and 30% of poultry is by-product. These by-products are rendered, ground and available as a feed source.

Animal protein meals are usually defined by inputs. Those specifically used in poultry diets include meat (no bone) or meat and bone meal from ruminants and/or swine; blood meal; poultry by-product meal; feather meal; and fish meal. There are specific limitations now assigned to these products with regards to inputs used and guarantees with respect to minimum nutrient levels. For example meat and bone meal may be specifically from ruminants and must be free of hair, wool and hide trimmings, except where it is naturally adhering to heads and hoofs. The products are rendered, which is a biosecure process that evaporates water, extracts fat and yields a finished ground product high in protein (which has no resemblance to the raw product) and minerals.

The products are marketed with guarantees as to minimum protein, phosphorus and calcium levels.

There are some challenges associated with the use of animal protein sources. First, food safety is the most important concern people have about the recycling of animal protein meals back through animals as feed ingredients. This is based on the links between the prion disease bovine spongiform encephalopathy (BSE – mad cow disease) and a variant Creutzfeldt-Jakob disease in humans. Importantly for poultry production though, researchers have been unable to demonstrate the transfer of prions to poultry (Moore J et.al. (2011) BMC Res Notes. Vol.4, p.501) and no symptoms of disease have been observed in birds up to five years after direct challenges. The proteins (prions) associated with BSE are not destroyed by traditional methods of rendering and are capable of causing disease when BSE contaminated meat and bone meals are injected cerebrally into ruminants.

As a consequence of the public's concerns about BSE, Australia does not allow the use of ruminant by-products in feed for ruminants; however, ruminant by-products are available for use in poultry feed.

Feedmill

In addition to BSE contamination, there are concerns that animal protein meals are responsible for food borne pathogen contamination, such as Salmonella. Typically these bacteria are destroyed by rendering and possible recontamination is often negated by pelleting of manufactured feeds. In most cases, if poultry acquire Salmonella it is likely to be from an environmental source other than feed. It is possible for animal protein meals to be contaminated with high levels of heavy metals, dioxins and PCBs (pesticides); however, meals are monitored and regulated to minimise this contamination.

Secondly, with respect to feeding the animal protein meals, the important practical issue is the variability in available nutrients (those that can be absorbed and retained by the bird) and limits to incorporation to maintain a diet balanced for all nutrients, particularly calcium and phosphorus. Table 3 shows the determined averages that are used in determining nutrient levels for meat and bone, blood, feather and poultry meals.

Table: ME values Nutrient levels in selected animal protein meals

Nutrient	Meat & Bone	Blood	Feather	Poultry
ME (MJ/kg)	11.2	15.2	13.7	13.1
Protein (%)	50.4	88.9	81.0	60.0
Fat (%)	10.0	1.0	7.0	13.0
Calcium (%)	10.3	0.4	0.3	3.0
Phosphorus (%)	5.1	0.3	0.5	1.7
Lysine (%)	2.6	7.1	2.3	3.1
Methionine (%)	0.7	0.6	0.6	1.0
Cystine (%)	0.7	0.5	4.3	1.0

Animal protein meals provide a good source of essential amino acids (e.g. lysine and methionine) and are also good sources of energy and minerals (particularly calcium and available phosphorus). However, there can be significant variation in availability (absorption and retention) of amino acids due to the day to day variation in inputs as well as processing conditions (temperature, moisture, pressure and time). The variation within processing plants can often be greater than variation between plants. It is important for users to establish strict criteria as to the quality of product and work with their suppliers to ensure these criteria are met. Quality should include measurements that indicate moisture; nutrient availability (particularly essential amino acids); levels of minerals (for example, calcium can vary from 8–12%; phosphorus from 4–6%); and stability of fat (all meals should be stabilised with an antioxidant).

The most accurate way of measuring the 'feed value' of an ingredient is to use an animal assay or bioassay. However, these assays are extremely time consuming and expensive. One of the most promising predictors of nutrient level and availability is near-infrared reflectance spectroscopy. This technology is rapidly being adopted by feed manufacturers and enables rapid screening of incoming products for a wide variety of measurements (moisture, protein, amino acid availability, fat, etc.). In most cases the samples can be prepared, scanned and results assessed in a few minutes. However, calibrations are still being established for meals and further research is required to classify the cause of variation in feed value.

Animal protein meals have a long history in poultry nutrition. Utilisation of this valuable feed ingredient is important in minimising loss (nutrient and economic value) in the production of safe, high quality poultry meat, eggs and bioproducts.

The typical ME values and nutrient composition of common animal protein sources are shown in Table.

Table: ME values and nutrient composition of selected animal protein sources

Ingredient	Protein (%)	ME (kcal/kg)	Calcium (%)	Available P (%)	Lysine (%)
Meat meal	50.0	2500	8.00	4.00	3.6
Fish meal	60.0	2720	6.50	3.50	5.3
Poultry by-product meal	60.0	2950	3.50	2.10	3.4
Blood meal	80.0	2690	0.28	0.28	6.9
Feather meal	85.0	3016	0.20	0.75	1.7

Fats and Oils

Fats and oils, collectedly termed lipids, are regularly used in poultry feed to satisfy the energy need of the animal as lipids have more than twice the amount of ME compared with carbohydrates or proteins per kg weight. Lipids are also an important carrier for fat soluble vitamins (A, D, E, and K) as wells for the provision of an essential fatty acid, linoleic acid, in the diet. A variety of fats and oils are used in feed, including lipids of animal origins (usually fats, i.e., tallow, lard, except fish oil) and lipids of vegetable origin (usually oils, i.e., soy oil, canola/rapeseed oil, sunflower oil, linseed oil, palm oil, cottonseed oil).

In practical feed formulation, the level of lipids rarely exceeds 4% in compound feed. However, even a small decrease in digestibility can cost dearly in terms of dietary energy. Like any other nutrient, a varying proportion of lipids are undigested depending on their sources and the species and age of the animal to which they are fed. Some of the data are summarised in Table.

Table: Lipid source and bird age on total tract digestibility of lipids

Lipid source	Digestibility (%)	Bird age (week)	Digestibility (%)
Tallow	73.6	1	53.2
Soy oil	85.0	2	80.7
Tallow-soy blend	75.4	3	85.9
Poultry fat	82.1	5	85.7
Palm oil	77.2	Average	76.4

It is surprising that nearly a quarter of dietary lipids are lost in the excreta of chickens. The significance of this can be seen from the fact that even with a seemingly small amount of inclusion, say 2.5% added fat in feed, it contributes as much as 7-9% of the dietary energy of a typical poultry diet. Thus, any improvement in digestibility, which may be achieved via the use of appropriate additives, such as enzymes, acidifiers and emulsifiers, will have a significant impact on the energy content of diets.

Minerals and Vitamins

Minerals are vital for normal growth and development in poultry, such as bone formation and body processes such as enzyme activation. Some minerals such calcium and phosphorus are required in large quantities. For example, laying hens require between 3.5-4% calcium, 0.3-0.4% available phosphorus and 0.2% sodium in their diets for egg production. Other minerals, such as copper, iron, manganese, zinc, selenium, cobalt, iodine and molybdenum, are required in milligram quantities but deficiency of these minerals will lead to serious health problems in mild cases and death in severe cases.

Similarly, vitamins are essential for the body systems of poultry. Both fat soluble (A, D, E, K) and water soluble (biotin, choline, folic acid, niacin, riboflavin, thiamine, pyridoxine, pantothenic acid and B12) are needed in the diet to maintain proper health and wellbeing of poultry.

Some vitamins and minerals are provided by most ingredients but the requirements for vitamins and minerals are generally met through premixes added to the diet. Diets may also contain additives for specific purposes.

Poultry Housing

Poultry is housed for comfort protection, efficient production and convenience of the poultry man.

Essentials of Good Housing

Comfort: The best egg production is secured from birds that are comfortable and happy. To be comfortable a house must provide adequate accommodation; be reasonably cool in summer, free-from draft and sufficiently warm during the winter provides adequate supply of fresh air and sunshine; and remain always dry. Given these the hen responds excellently.

Protection: Includes safeguards against theft and attack from natural enemies of the birds such as the fox, dog, cat kite, crow, snake, etc. The birds also should be protected against external parasites like ticks, lice and mites.

Convenience: The house should be located at a convenient place, and the equipment so arranged as to allow cleaning and other necessary operations as required.

Location of Poultry House

In planning a poultry house, the location should be taken into consideration. In selecting site for poultry houses the following factors should be considered.

- Relation to other building: The poultry house should not be close to the home as to create unsanitary conditions. On the other hand it should not be too far away either because this will require more time in going to and for in caring for the birds. In general at least three trips should be made daily to the poultry house in feeding, watering, gathering the eggs, etc.

- Exposure: The poultry house should face south or east in moist localities. A southern exposure permits more sunlight in the house than any of the other possible exposures. An eastern exposure is almost as good as a southern one. Birds prefer morning sunlight to that of the afternoon. The birds are more active in the morning and will spend more time in the sunlight.

- Soil and drainage: If possible the poultry house should be placed on a sloping hillside rather than a hilltop or in the bottom of a valley. A sloping hillside provides good drainage and affords some protection. The type of soil is important if the birds are to be given a range. A fertile well drained soil is desired. This will be a sandy loam rather than a heavy clay soil. A fertile soil will grow good vegetation which is one of the main reasons for providing range. If the poultry house is located on flat poorly drained soil, the yards should be tiled otherwise the birds should be kept in total confinement.

- Shade and Protection: Shade and protection of the poultry house are just as desirable as for the home. Trees serve as a windbreak in the winter and for shade in the summer. They should be tall, with no low limbs. Low shrubbery is no good as in their presence the soil becomes contaminated under the shrubbery, remains damp/ and sunlight cannot reach it to destroy the di ease germs. One thing we should remember that plenty of sun shines should be available at the site.

Housing Requirements

Floor space: The smaller the house the more square feet are required for each hen.

Bigger pens have more actual usable floor space per bird than smaller pens. The recommend at as suggested might be useful regarding floor, feeders and watering space.

For economic production of laying hens it is always better to keep them in small unit of 15 to 25 birds. This number can go up to a maximum limit of 250 birds. In commercial poultry farms units of 125 or so are advisable. Where there is a long house, partitioning at every 20 feet should be made to eliminate drafts, etc.

Ventilation: Ventilation in the poultry house is necessary to provide the birds with fresh air and to carry off moisture. Since the fowl is a small animal with a rapid metabolism its air requirements per unit of body is high in comparison with that of other animals. A hen weighing 2 kg and on full feed, produces about 52 liters of CO_2 every 24 hours. Since CO_2 content of expired air is about 3.5 per cent, total air breathed amounts to 0.5 liter per kg live weight per minute. A house that is a tall enough for the attendant to more around comfortably will supply far more air space than will be required by the bird's that can be accommodated in the given floor space.

Temperature: Hens need a moderate temperature of 50°F to 70°F. Birds need warmer temperature at night, when they are inactive, than during the day. The use of insulation with straw pack or other materials, not only keeps the house. Warmer during the winter months but cooler during the summer months Cross ventilation also aids in keeping the house comfortable during hot weather.

Dryness: Absolute dry conditions inside a poultry house is always ideal condition dampness causes discomfort to the birds and also gives rise to the diseases like colds, pneumonic etc. Dampness in poultry house caused by: (1) moisture rising through the floor; (2) leaky roofs or walls; (3) rain or snow entering through the windows; (4) leaky water containers; (5) exhalation of birds.

Light: Daylight in the house is desirable for the comfort of the birds. They seem more contented on bright sunny days than in dark, cloudy weather. Sunlight in the poultry house is desirable not only because of the destruction of disease germs and for supplying vitamin-D but also because it brightens the house and makes the birds happy. Birds do fairly well when kept under artificial lights.

Sanitations: The worst enemies of the birds, i.e., lice, ticks, fleas and mites are abundant in poultry houses. They not only transmit diseases but also retard growth and laying capacity. The design of the house should be such which admits easy cleaning and spraying. There should be minimum cracks and crevices. Angle irons for the frame and cement asbestos or metal sheets for the roof and walls are ideal construction materials, as they permit effective disinfection of the house. When wood is to be used, every piece should be treated with coaltar, cresol, or similar strong insecticides before being fitted.

Poultry Housing Systems

Poultry housing should have proper accessibility, safety, exterior appearance, and appropriateness of design are important aspects of housing for your poultry flock.Before begin to build poultry housing system you should consider how you will access and maintain your poultry housing. You should choose a design that allows for easy access to nests, perches, feeders, and water. Suitable access will make it easier to clean all part of the coop.

Before designing, building, and maintaining your coop, take action to prevent possible injury to you or your birds. You should remove any loose or ragged wire, nails, or other sharp-edged objects from the coop. You should make ensure that the birds can perch on only roosts that you are providing. You should remove access to other perching areas, eg. windowsills, nest box tops, or electric cords, whenever possible.

If the poultry house is visible to your neighbours, you make ensure that it does not detract from the overall appearance of its surroundings. You can improve the looks of your poultry coop by painting and properly maintaining the exterior. Removing weeds and trash from around the coop not only enhances its appearance but also helps with rodent control. you should do landscaping that can screen your poultry coop from neighbours as well as help muffle the sounds your flock produces.

Before making poultry housing system one should keep following things in mind:

- The house should protect birds from adverse climatic conditions.

- It should have easy and economic operation.

- You can do scientific feeding in a controlled manner.

- Provide proper micro-climatic conditions in a near vicinity of bird.

- Should have an effective disease control measures.

- You can do proper supervision.

Before selection of location of poultry housing system:

- Your poultry house shouldn't be near residential and industrial area.

- It should have proper access with road facilities.

- It should have the basic amenities like water, feed and chicks transportation and electricity.

- Wherever you can find farm labourers at relatively cheaper wages.

- It should be located in an elevated area and should not be any water-logging.

- It should have proper ventilation and should be in open area.

Before making layout preparation:

- Layout should not allow visitors or outsiders vehicles near the birds.

- The sheds should be ideally located that the fresh air first passes through the brooder shed, followed by grower and layer sheds. This will prevent the spread of diseases from layer houses to brooder house.

- There should be from a minimum distance of 50-100 feet between chick and grower shed and the distance between grower and layer sheds should be of minimum 100 metre.

- The egg store room, office and the feed store room should be located near entrance to overcome the movement of people around the poultry sheds.

- The disposal pit and sick room should be constructed only at the extreme end of the site.

For poultry housing another important factor to consider is the poultry housing system and fencing options. Poultry housing system determines around 40% of the rate of success and profitability of your poultry farming business. A good poultry housing system is very important for the success of your poultry faming.

Following poultry housing system is generally commonly used in poultry farming:

- Extensive Housing System: Range and fold unit.

- Semi Intensive Housing System: Standard semi intensive unit and straw yard.

- Intensive Housing System: 1. Deep Litter 2. Wire & Slated Floor 3. Straw Yard 4. Battery Cage.

A. Free Range System:- Under free range poultry farming system, birds are roaming around on their own in search for food. This poultry housing system cannot be used for commercial poultry farming because of high level of risks involved in it.

B. Deep Litter System:- This system involves spreading wood shavings or saw dust on a concrete ground with the chickens placed on the floor and the dust serving as a 'cushion' for them. The dust is changed regularly, either by removing and replacing or by layering. This system is used by a lot of poultry farmers because it increases efficiency and it is easier to manage large flocks of birds but it also allows to spread of diseases when there is an outbreak and also makes it difficult to fish out unproductive birds.

C. Battery Cage System:- As for the battery cage housing system, metal cages are made inside a building to house the birds. The cages are generally split into different compartments to keep small groups of birds. The cages also contain feeding and water pots as well as laying nests. Bird's droppings fall on the floor and it is cleaned mechanically by a scraper. This method is considered to be the most effective for a layer farming but the major downside of this method is that it is more expensive to construct and maintain in compare to other types of poultry housing systems.

Intensive housing system is the best housing, if you are running a poultry farm for commercial purposes and you intend getting the best out of the business, in term of high birds' productivity and efficiency. So you should consider for intensive housing system.

There are two important housing systems in poultry farms. They are:

- Deep Litter System
- Cage System

1. Deep Litter Poultry System

This is the conventional housing system of poultry farms practiced since long back. The floor in the poultry farm is tube covered with a layer of saw dust, paddy husk or straw. This layer is about 4 inches in thickness in summer, while in other seasons 7 inches thick layer is used on the floor. This layer of saw dust, paddy husk or straw which covers the floor of the poultry farm is called litter. The litter should be always in dry condition. The droppings of birds produce ammonia. In wet condition, more ammonia in the litter may cause diseases. Hence, it should be mixed periodically with lime to reduce the percentage of ammonia in the litter. Further the litter should be changed once in a year. Growing chicken on floor covered by litter is said to be the Deep litter system (deep litter) or loose housing system.

The dimensions of the poultry house depend on the number of chicks to be accommodated. For example, 125 chicks require a space of about 400 square feet. A chick normally requires 2.25 square feet. It should be remembered that more than 250 chicks should not be accommodated in a poultry house. A distance of 40 feet is to be maintained in between two poultry houses or sheds.

2. Cage System in Poultry

Of late, many farmers are growing chicks by using cage system. In this system, the birds are grown in cages made of iron mesh. For growing about 25 fowls a cage measuring is 36" x 48" is used. Lesser dimension cages are used for growing small flocks. The cages are arranged in 2 or 3 tiers in poultry houses. Feeding, watering and other facilities are given to the chicken introduced in these cages.

Advantages of Cage Method in Poultry

1. In this method, there is no wastage of space and food.
2. Chicks involved in cannibalism may be easily identified and removed. Cannibalism in poultry is referred to as one bird pecking at the other bird.
3. The diseased birds can easily be separated.
4. Less labor is involved in this method.
5. Collection of eggs is an easy process.
6. Mortality rate is less in this system.

Disadvantages of Cage System in Poultry

1. The cracked eggs are common.

2. It is difficult to clean the individual cages and to dispose the manure.

3. The food may not be distributed accurately depending on the consumption of the birds in the cage.

4. The birds are frequently subjected to a disease called cage layer fatigue. The bird lies on its side giving an impression that it is affected with paralysis. Hence, this is referred to as cage paralysis or cage layer fatigue.

5. Fatty liver syndrome is common in cage system. It is characterised by

    (a) a gradual drop in egg production (40% - 10%),

    (b) fatty, enlarged, tan colored liver is noticeable, and

    (c) birds become fatty.

In cage culture certain precautions are to be followed. They are:

1. In summer, water is to be sprinkled on cages. This is due to the fact that cage poultry needs protection from excessive heat in severe summer.

2. The chicks should be subjected to dubbing and debeaking. As a result of de-beaking, the chicken may not involve in cannibalism and can easily procure the soft food without waste.

Feather Pecking

To some people chickens are just those brown feathered things that lay the eggs you keep in the fridge. To others (the special, initiated, fans of chickens) they are a constant source of enjoyment and as much a member of the family as any other pet.

However, whatever your view on chickens, they have become embedded in the very lexicon of our language. If you are in a small space you feel "cooped up". Perhaps your "chickens will be coming home to roost" because you have "chickened out" and are now "running around like a headless chicken".

Possibly the most pervading though has to be "the pecking order", from the Houses of Parliament to Poultry Houses there are pecking orders, however, only in the latter case does any literal pecking occur.

Identifying Feather Pecking

Chickens will always peck at each other a little bit, this is almost an essential part of establishing a "pecking order" or hierarchy amongst the flock. So the 'top hen' will peck at others and the lowest bird in the order will be pecked most.

Most of the time this isn't a real problem and you'll probably find that the bird at the bottom of the social order will merely high tail it away if it gets fed up and no real damage is done. However, if your hens are kept in an enclosed run, simply running away might not be an option.

Feather pecking is, as the name suggests, when one hen starts pecking at another's feathers and pulls them out causing distress to the birds and, in some cases, draws blood from a wound.

This is not to be confused with an over-amorous cockerel who can often pull feathers out of a hens back while trying to gain a perch, or even the annual moult. Feather pecking often occurs around the hen's vent, tail or head. A good indication that feather pecking is occurring is not only bald patches on the birds but the sign of half eaten feathers around the coop.

Chickens are attracted to the colour red (one reason why so many poultry feeders and drinkers are red) so the comb, wattle and the vent - which is often reddened after laying an egg - draws the attention of other hens. Chickens are also attracted to blood, the colour and smell (they are after all essentially mini dinosaurs) so, if wounds are left untreated, it will only attract them more. Always make sure you treat any blood or wounds straight away with either a gentian violet spray or a wound powder, for the well-being of the bird.

Boredom

This is number one on the list. If your hens have to be kept in an enclosed space (for their safety and your sanity) then perhaps this is the easiest to solve. Convention says that a hen needs 3 sq. ft. of outdoor space when permanently enclosed in a run. You might find giving your birds that little extra bit of space, if possible, will solve the problem.

Equally, like all living things, chickens like to feel the sun on their backs (but they will still need shade options too). If you have them tucked away in a dark and dingy corner of the garden, can you move them so that they can enjoy more daylight? If neither of these are an option for you, giving the birds something to occupy themselves is one way to cure boredom.

Hanging Boredom Buster Chicken Treats or perhaps a few fresh greens hung around the run will help keep your hens distracted and busy. Do your hens have a nice friable surface to scratch around in during the day to look for bugs and worms e.g. woodchip? Can you give them some perches in their run perhaps, or branches to stand on?

Stress

Stress is also one of the top reasons for feather pecking. A chicken coop that is too hot will stress the birds, try to increase the ventilation to cure this problem. If you have too many hens in one hen house or poultry run, the lack of space will stress them. Can you reduce the overcrowding? As a guide a hen should be allowed 1 to 2 sq ft in their house with 7"-8" of perch space to roost.

When laying an egg especially, chickens like the area to be dim and quiet, excessive or glaring lights will only stress them and reduce egg laying. A chicken coop should have at least one nest space per 4 to 5 birds.

Anti-Feather Pecking Sprays can help to break a feather pecking habit such as this. The spray is applied to the victim bird, it creates a foul taste in the perpetrators mouth, helping to deter them from pecking. Some pecking sprays are clear liquids and contain antiseptics, whilst others are brown and based on Stockholm Tar which is an old traditional method to stop pecking or biting. You may find one type will work with your flock whilst another may not so you may need some trial and error to find the right one.

The Annual Moult

For laying hens the moult is a yearly process, usually around the end of summer and the start of autumn, in which a chicken replaces its feathers over a period of approx. a month. The moult can begin, surprisingly, as late as early winter in some birds.

They will look really bedraggled and 'under the weather' during this time, and therefore enticing to other hens, who will peck at the newly emerging quills and the area of reddened skin. This can quickly escalate to cannibalism so should be dealt with straight away.

Gentian Violet Spray will work as both an antiseptic and colour the affected area purple. As chickens are attracted to the red skin of a pecked bird, turning the skin purple is an easy solution to helping the poor bird. An anti-feather pecking spray would help in this instance also, making the newly emerging feathers taste bad to the pecking hens.

To help them through the yearly moult you might try the traditional method of putting some cod liver oil in their feed, at a rate of around a tea spoon per Kg of feed. Rich in vitamins A and D it helps with the general conditioning of your bird, especially useful around the moult. Other tips are to give them high protein treats that will encourage speedy growth of feathers. A good quality Poultry Tonic or Verm-X Poultry Zest Pellets would also be advisable at this time so their body can recover quickly.

Lack of Protein

A lack of protein in the diet is a rarer reason for feather pecking, but it is a consideration if you see your birds pulling out and eating the feathers but don't feel that this is in an aggressive manner.

Feathers are made of protein, so if a bird is suffering a protein deficiency they can turn to feather pecking and eating the feathers as a source for much needed protein. Making sure that your hens have a good, balanced diet, in particular a quality layers pellet as their staple feed, will help keep this to a minimum.

Their Layers Pellets should contain around 16% protein, this will be on the label e.g. the Smallholder Layers Pellets range, or, Garvo Alfamix for Chickens.

Until the recent introduction of an EU law which said that we can no long sell (market) mealworms as being 'for chickens', dried mealworms would have been suggested as a good source of protein of course, as they are a high protein food, but they are now only available in our wild bird feed section, Dried Mealworms for Birds.

Persistent Peckers

If you have a persistent pecking problem and none of the above solve the problem, then the answer might be segregation or isolation, for either the bird who is doing the pecking or perhaps the bird who is being pecked, to give her time to recover and grow new feathers safely.

When a bird is segregated it should still be located near to the other hens, where it can be seen by the remaining flock. This is an important point. While hens can still see each other, they remain part of the same flock, even if separated. If you completely remove the hen and then try to re-introduce it, you might find feather pecking starts all over again as they bully the 'new' member of the flock. Equally, if you're separating the culprit from the flock, leave them apart from the others for around four days or so, at which point they will have hopefully forgotten their aggressive behaviour.

Pinless Peepers

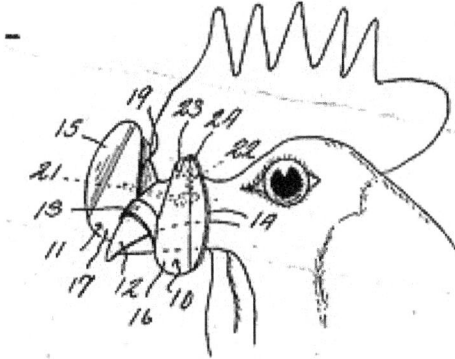

From the U.S. Patent *"Device to prevent picking in poultry"* filed

Blinders, also known as peepers, are devices fitted to, or through, the beaks of poultry to block their forward vision and assist in the control of feather pecking, cannibalism and sometimes egg-eating. A patent for the devices was filed as early as 1935. They are used primarily for game birds, pheasant and quail, but also for turkeys and laying hens. Blinders are opaque and prevent forward vision, unlike similar devices called spectacles which have transparent lenses. Blinders work by reducing the accuracy of pecking at the feathers or body of another bird, rather than spectacles which have coloured lenses and allow the bird to see forwards but alter the perceived colour, particularly of blood. Blinders are held in position with a circlip arrangement or lugs into the nares of the bird, or a pin which pierces through the nasal septum. They can be made of metal (aluminium), neoprene or plastic, and are often brightly coloured making it easy to identify birds which have lost the device. Some versions have a hole in the centre of each of the blinders, thereby allowing restricted forward vision.

Benefits

In pheasants, blinders have been shown to reduce the incidence of birds pecking each other and damage to the feathers or skin.

In laying hens, blinders have been shown to reduce feather pecking, improve food utilisation (due to less spillage) and increase egg production.

Welfare Concerns

Blinders which require a pin to pierce the nasal septum and hold the device in position almost certainly cause pain to the bird. In the UK, the use of these devices is illegal on welfare grounds. The Department for Environment, Food and Rural Affairs in their Codes of Recommendations for the Welfare of Livestock: Laying Hens, states: "The Welfare of Livestock (Prohibited Operations) Regulations 1982 (S.I. 1982 No.1884) prohibits the fitting of any appliance which has the object or effect of limiting vision to a bird by a method involving the penetration or other mutilation of the nasal septum."

Studies on pin-less blinders indicate these devices are at least temporarily distressful to the birds. In pheasants, fitting blinders causes an increase in head shaking and scratching, and increases in damage to the beak and nostrils of the bird. Fitting pin-less blinders to laying hens leads to reduced activity, increased resting, adjustment problems in feeding, stereotypic head shaking and protracted displacement neck preening for a month after fitting. In another study on laying hens, mortality was greater among hens wearing blinders compared to hens that had been beak-trimmed.

Debeaking

Beak trimming in poultry management is the act of reducing the length of the beak of poultry birds. The purpose of doing this is to prevent feather pulling and cannibalism and to reduce feed wastage. It is a delicate operation, and if it is improperly done, it may leads to difficulties in drinking and eating, which directly leads to poor growth, unevenness in flock and even mortality as a result blood loss.

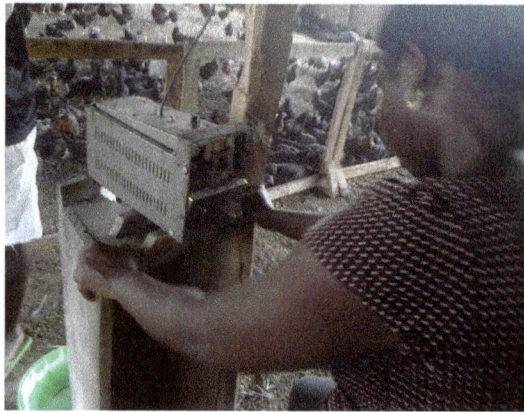

The operation can be carried out at one week-old (7-9 days) and few weeks-old (8-10 weeks). The advantage of debeaking at one-week-old is that, the operation would have a minimum effect on chick's body weight and it is not necessary to carry it out again a second time during the rearing period. For precise beak trimming operation, the birds should be few weeks old (8-10 weeks). The general disadvantage of debeaking or beak trimming is that, when it is improperly done, it could take long for the birds to regain body weight.

Costs

The costs of beak trimming relate primarily to welfare concerns. These include acute stress, and acute, possibly chronic, pain following trimming. A bird's ability to consume feed is impaired following beak trimming because of the new beak shape and pain. Most studies report reduced body weights and feed intake following beak trimming, however, by sexual maturity or peak egg production, growth rates are usually normal. Weight losses were reduced in chicks that were beak trimmed by infrared compared with chicks trimmed by a hot-blade.

The Pain of Beak Trimming

Whether beak trimming causes pain is a hotly debated concern. It is a complex issue as it may involve acute and/or chronic pain, and depends on the age it is performed, the method of trimming and the length of beak that is removed. Beak trimming in the poultry industry usually occurs without anaesthetic at 1-day of age or when the chicks are very young, but can occur at a later age if an outbreak of feather pecking occurs, and in some cases, birds may be beak trimmed on repeated occasions. Beak trimming is not permitted in the UK on meat chickens that are aged over 10 days.

White Leghorn pullets showing the results of beak trimming

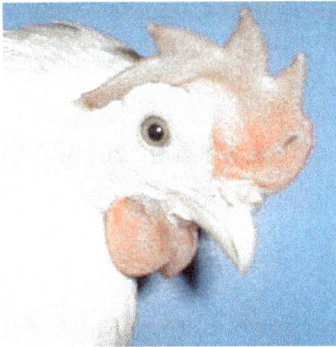

Non-beak trimmed

Beak trimmed

Acute Pain

The beak is a complex, functional organ with an extensive nervous supply including nociceptors that sense pain and noxious stimuli. These would almost certainly be stimulated during beak trimming, indicating strongly that acute pain would be experienced. Behavioural evidence of pain after beak trimming in layer hen chicks has been based on the observed reduction in pecking behavior, reduced activity and social behavior, and increased sleep duration. In Japanese quail, beak-trimming by cauterization caused lower body weights and feed intake in the period just after beak trimming. Beak trimmed Muscovy ducks spent less time engaging in beak-related behaviours (preening, feeding, drinking, exploratory pecking) and more time resting than non-trimmed ducks in the days immediately post-trim. These differences disappeared by 1 week post-trim. At 1 week post-trim the trimmed ducks weighed less than non-trimmed ducks, but this difference disappeared by 2 weeks post-trim. It is, however, unclear if the above changes in behaviour arise from pain or from a loss of sensitivity in the beak. Pecking force has been found to decrease after beak trimming in adult hens possibly indicating that hens are protecting a painful area from further stimulation. However, pecking force did not differ between chicks with or without minor beak-trims at 2 to 9 days of age, suggesting that chicks with minor beak-trims do not experience pain from the beak.

Chronic Pain

Severe beak trimming, or beak trimming birds at an older age is thought to cause chronic pain. Following beak trimming of older or adult hens, the nociceptors in the beak stump show abnormal patterns of neural discharge, which indicate acute pain. Neuromas, tangled masses of swollen regenerating axon sprouts, are found in the healed stumps of birds beak trimmed at 5 weeks of age or older and in severely beak trimmed birds. Neuromas have been associated with phantom pain in human amputees and have therefore been linked to chronic pain in beak trimmed birds. If beak trimming is severe because of improper procedure or done in older birds, the neuromas will persist which suggests that beak trimmed older birds experience chronic pain, although this has been debated.

Age and Methods of Debeaking

There are several methods of debeaking and several ages when birds can be debeaked. The age of debeaking will determine the procedure.

1. Block Debeaking at 5 to 7 Days

This is the best procedure for pullets to be used for egg production. Most of the farmers in Nigeria are sceptical about debeaking their birds at this age. This operation can be compared to the cutting of umbilical card in new born babies. Earlier you do this operation lesser the stress and pain.

The chicken are easy to handle at this age and the debeaking process is faster. This precision debeaking needs considerable experience to do a good job. When the beak is cut and cauterized correctly it will not grow back and there is no need for a later touch up. The farmers can utilize the services of the hatcheries to take up this operation.

The chicks are debeaked with an electric debeaking machine having a guide plate with hole. A guide with a larger hole should be used if the chicks are larger than normal. The chick is held with the thumb on the back of the head and forefinger under the throat. The closed beak is inserted in the hole and light pressure exerted on the throat to pull back the tongue. The beak hits the trigger which allows the hot blade drop down and automatically make the cut.

The lower beak should be slightly longer than the upper. This can achieved by tilting the chick's head downwards at the time the beak is inserted in the hole. The severed beak must be kept in contact with the blade for exactly 2 seconds. When held longer, stress is too great and if shorter, the beak will grow back.

2. Debeaking between 2 to 10 Weeks

Many farmers wait until the birds are 5 or 6 weeks of age or until they see the first traces of cannibalism in the flock before they debeak. This require a different procedure. However debeaking after 8 days of age creates a severe stress. Pullets are vulnerable for IBD outbreak between 3 to 6 weeks of age. Any stress at this age may lead to outbreak.

Electric debeaking machine can be used for this purpose. It is advisable to remove one-third to one-half of the beak. The lower beak should remain longer than the upper one. With 10 to 12 weeks and above old birds, it will be necessary to cut one beak at a time.

3. 16 to 18 Week Debeaking

As a last resort, birds may be debeaked around this age. But do not remove as much of the beak. Debeak each mandible separately and the lower beak of pullets should be at least 0.3cm longer than the upper beak to avoid feed wastage.

4. Touch-up Debeaking

If the early debeaking has not been done properly, many beaks will partially grow back by the time the pullets are 8 to 9 weeks of age or older and will need to be 'touched-up' by another partial debeaking. This is a common procedure.

Points to Observe during Debeaking

1) A blade temperature of between 700-800 °C is sufficient.

2) The knife should be "cherry" red. Too hot or not hot enough gives bad results.

3) Do not de-beak under extreme temperatures. The best temperature is 20- 25 °C.

4) Remove the feed 4- 5 hours before de-beaking and give feed immediately after de-beaking. When the birds are hungry, they will start eating directly afterwards; this stops possible bleeding.

5) Supply feed and water adequately immediately after debeaking.

6) Change the blades regularly. Do not use old blades.

7) Always replace the new knife in the proper way, the straight side of the edge in front of you and the sloping side on the backside.

8) Use vitamin-K and broad spectrum antibiotics like Oxytetracyclines in water, 1 day before and 3 days afterwards.

9) It is recommended to check the beaks before production starts. Birds with too long beaks or very long lower beaks should be trimmed again.

10) Prevent the piling of chicks during the debeaking operation.

11) During the debeaking keep the tongue away with the index finger, between the upper and lower beak, to avoid burning.

12) Do not debeak during a vaccination period.

13) Never use sulfa drugs during debeaking. This may increase the bleeding.

Advantages and Disadvantages of Debeaking.

There are advantages and disadvantages to debeaking. But certainly the advantages far outweigh the disadvantages.

Advantages

1. Pecking is reduced.
2. Helps in preventing feather picking and cannibalism.
3. Feed efficiency is improved.
4. Liveability is better.
5. Lesser number of culls.
6. Uniformity of the flock is better.

Disadvantages

1. Birds lose weight for 1 to 2 weeks after debeaking.
2. Growth rate is reduced after debeaking.
3. Late debeaking may slightly delay sexual maturity.

Dubbing

Dubbing is the procedure of removing the comb, wattles and sometimes earlobes of poultry. Removing the wattles is sometimes called "dewattling".

A cockerel that appears to have been dubbed

A cockerel that has not been dubbed showing the comb on top of the head, the wattles hanging from the throat and the earlobes (pinky white) hanging from the ears

Procedure

To perform dubbing, the tissues are first disinfected and, if available, an anesthetic is applied to limit pain. Sterile scissors or dubbing shears are used to cut the tissues off, and a styptic, an astringent chemical that reduces bleeding, is applied. The wounds are left uncovered. Some recommend dubbing should be done on day old chicks whilst others advise waiting until the bird's comb is more developed.

Benefits

Dubbing is sometimes performed to limit damage caused by injury or frostbite. Dubbing for some breeds has become a tradition and is required for some birds to meet breed-specifications. Other reasons include removing combs which have become so large they prevent the bird from taking food into its mouth or making the head so heavy it sinks into the bird's chest. In the US, the National Chicken Council (2003) listed dubbing of cockerels as one of the acceptable procedures that may cause short-term stress but which are necessary for the long-term welfare of the flock. Dubbing is also performed to prevent injuries from other birds or while being kept in pens.

References

- Marchant-Forde, R.M. and Cheng, H.W., (2011). Different effects of infrared and one-half hot blade beak trimming on beak topography and growth. Poultry Science, 89: 2559-2564. doi:10.3382/ps.2010-00890

- "FAWC report on broiler breeds". Archived from the originalon 17 June 2014. Retrieved 5 December 2011

- "EFSA (2010). Scientific opinion on welfare aspects of the management and housing of the grandparent and parent stocks raised and kept for breeding purposes". EFSA Journal. 8 (7): 1667. doi:10.2903/j.efsa.2010.1667. Retrieved 9 December 2011

- Breward, L. and Gentle, M.J., (1985). Neuroma formation and abnormal afferent nerve discharges after partial break amputation (beak trimming) in poultry. Experientia, 41: 1132-1134. doi:10.1007/BF01951693

- Jones, E.K.M. and Prescott, N.B., (2000). Visual cues used in the choice of mate by fowl and their potential importance for the breeder industry. World's Poultry Science Journal, 56: 127-138. doi:10.1079/WPS20000010

Chapter 4

Cattle Management

Cattle are raised for meat, milk and hides. Some routine cattle husbandry practices are cleaning, feeding and milking. This chapter has been carefully written to provide an easy understanding of important cattle farming practices such as automatic milking, milk separation, use of milk cooling tanks and Bovine Somatotropin, etc.

Cattle Farming

Cattle farming involves rearing and management of two types of animals- one group for food requirements like milk and another for labor purposes like plowing, irrigation, etc. Animals which provide milk are called milch/dairy animals. For example, goats, buffalo, cows, etc. Animals which are used for labor are called draught animals.

Since dairy animals are cared and bred for milk, we need to improve the milk production to meet the requirements. The period after the birth of a calf, when a cow starts to produce milk, is called lactation period. We can enhance the milk production by increasing this lactation period. But along with milk production, quality must also meet. Dairy farm management is the management of the milch animals with the goal of enhancing the quantity and quality of the milk produced. For this reason, high yielding and disease resistant breeds are developed.

For example, the foreign breeds like Jersey, Holstein-Friesian, Brown Swiss, have long lactation period while local breeds like Red Sindhi, Sahiwal, and Gir are known for their disease-resistant trait. The breeding of these two varieties helped us to enhance the quantity and quality of the milk produced.

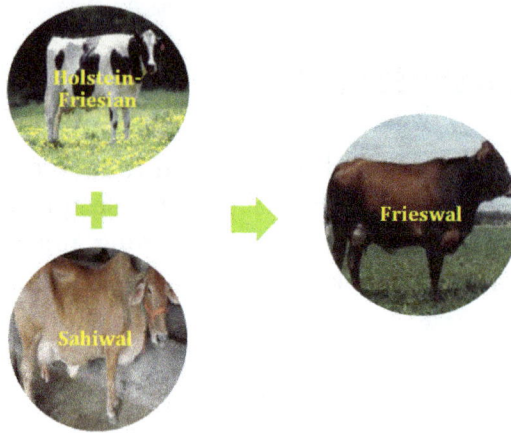

Cross-breeding in Cattle Farming

Farm Management

Cattle farming are not all about milk or meat production. It has some responsibilities to do. To meet the human requirements, we need to take good care and need proper management of livestock.

Shelter

In cattle farming, animals are maintained in a strictly hygienic manner with proper housing. For the maintenance, we need to follow some routines.

- Animals and their sheds need to be cleaned at a regular interval.

- Animals should be brushed regularly to remove the dirt and bugs in their body.

- The shed should be well-ventilated and roofed so that animals are protected from rain, heat, and cold.

- A proper drainage system should be there to remove animal waste.

Food

Food requirements of animals are also a part of cattle farming. To maintain high yielding and disease-resistant breeds, they must be provided with adequate water supply and nutrient-rich fodder regularly according to their needs. In cattle farming, two factors are considered regarding the food of cattle. The food that is provided must keep the animal healthy as well as it should meet the farming requirement. Hence, the animal feed includes roughage (high fiber content) and concentrates (high proteins and nutrient content). In addition to this, supplements containing micronutrients are also provided to animals. An adequate proportion of these rations promote healthy and high output animals.

Disease Management

The third responsibility of cattle farming management is to maintain disease-free breeds. Animals are not an exception to disease. They also suffer from numerous diseases. This may affect the health as well as productivity of animals; even cause their death. Parasites, bacteria, and viruses are the major villains here. These microbes infect the cattle externally as well as internally. Vaccination is the one solution for the protection against bacterial and viral infections.

Cattle Production

Cattle production involves keeping bulls and cows for meat (beef) or milk (dairy) purpose. Cattle feed on grasses, legumes, roughage etc. That is why they are called herbivores and they are also known as ruminants (because they have one stomach with four compartments). In Nigeria (West Africa), commercial beef cattle production is very common especially in the Northern part of the country.

Cattle production is the management of bulls and cows for production beef and/or milk purpose. Commercial Cattle production is where the goal of production is to make money through selling of live cattle or animal products (meat, milk, etc.). This enterprise is divided into three sub-production types, based on the purpose of production.

- Beef cattle production- for meat (beef) production only.

- Dairy cattle production- for milk production only.

- Dual-purpose production- for both milk and meat production.

Just like any other business enterprise, you need to carry out feasibility study before going to cattle production business. Before anyone can start cattle business, he or she must be sure that feeds and herbage are sufficiently available for the cattle. Cattle are herbivores i.e. they eat vegetables. Although, they can also take concentrates. Another thing about this business is that location and climate has a great influence on the performance of both bull (male cattle) and cow (female cattle). For example, cattle that have their native home in a temperate region may not survive or thrive in a tropical region. Therefore, before starting a beef cattle production business, the following points must be noted.

Facts about Cattle Farming

- It requires a lot of capital at the start and less after.

- It is a labor-intensive business.

- It could be a long term business if well-managed.

- The breed of cattle to be selected for beef production should have standard body conformation for beef cattle.

- Means of feeding and supplying water to the cattle should be considered and known.

- The prevalence of tsetse flies in the proposed site or location of production.

- 'De-horning' should be done for cattle with horns for easy handling and to prevent injury.

- Do not purchase animals from the same place.

- If there is enough land, establishment of pasture should be done before bringing in animals.

- Diseased animals can be a threat to other animals; so get animals that are certified healthy.

- Provide shade, good water and fence the farm round with barbed wire to prevent uncontrolled movement of animals.

- Breeding plans or program should be known beforehand.

- Consult experts in the field.

Dairy Cattle Farming

Dairy farms have grown significantly from their humble beginnings to meet the growing demand for dairy products. Although the size of the farms has grown they are still family farms. According to the U.S. Department of Agriculture, 98% of U.S. dairy farms are still family owned and operated. Sometimes these farms are still owned by relatives and descendants of the original owners. Whether the farm is large or small, farmers are committed to producing the highest quality product while taking care of the land they farm.

A Holstein cow with prominent udder and less muscle than is typical of beef breeds

Dairy cattle (also called dairy cows) are cattle cows bred for the ability to produce large quantities of milk, from which dairy products are made. Dairy cows generally are of the species Bos taurus.

Historically, there was little distinction between dairy cattle and beef cattle, with the same stock often being used for both meat and milk production. Today, the bovine industry is more specialized and most dairy cattle have been bred to produce large volumes of milk.

The United States dairy herd produced 84.2 billion kilograms (185.7 billion pounds) of milk in 2007, up from 52.9 billion kilograms (116.6 billion pounds) in 1950, yet there were only about 9 million cows on U.S. dairy farms—about 13 million fewer than there were in 1950. The top breed of dairy cow within Canada's national herd category is Holstein, taking up 93% of the dairy cow population, have an annual production rate of 10,257 kilograms (22,613 pounds) of milk per cow that contains 3.9% butter fat and 3.2% protein.

Management

Dairy cows may be found either in herds or dairy farms where dairy farmers own, manage, care for, and collect milk from them, or on commercial farms. Herd sizes vary around the world depending on landholding culture and social structure. The United States has 9 million cows in 75,000 dairy herds, with an average herd size of 120 cows. The number of small herds is falling rapidly with the 3,100 herds with over 500 cows producing 51% of U.S. milk in 2007. The United Kingdom dairy herd overall has nearly 1.5 million cows, with about 100 head reported on an average farm. In New Zealand, the average herd has more than 375 cows, while in Australia, there are approximately 220 cows in the average herd.

Dairy farming, like many other livestock raring, can be split into intensive and extensive management systems.

Cows on a dairy farm in Maryland, U.S.

Intensive systems focus towards maximum production per cow in the herd. This involve formulating their diet to provide ideal nutrition and housing the cows in a confinement system such as free stall or tie stall. These cows are housed indoors throughout their lactation and may be put to pasture during their 60-day dry period before ideally calving again. Free stall style barns involve cattle loosely housed where they can have free access to feed, water, and stalls but are moved to another part of the barn to be milked multiple times a day. In a tie stall system, the milking units are brought to the cows during each milking. These cattle are tethered within their stalls with free access to water and feed are provided. In extensive systems, cattle are mainly outside on pasture for most of their lives. These cattle are generally lower in milk production and are herded multiple times daily to be milked. The systems used greatly depends on the climate and available land of the region of which the farm is situated.

To maintain lactation, a dairy cow must be bred and produce calves. Depending on market conditions, the cow may be bred with a "dairy bull" or a "beef bull." Female calves (heifers) with dairy breeding may be kept as replacement cows for the dairy herd. If a replacement cow turns out to be a substandard producer of milk, she then goes to market and can be slaughtered for beef. Male calves can either be used later as a breeding bull or sold and used for veal or beef. Dairy farmers usually begin breeding or artificially inseminating heifers around 13 months of age. A cow's gestation period is approximately nine months. Newborn calves are removed from their mothers quickly, usually within three days, as the mother/calf bond intensifies over time and delayed separation can cause extreme stress on both cow and calf.

Domestic cows can live to 20 years; however, those raised for dairy rarely live that long, as the average cow is removed from the dairy herd around age four and marketed for beef. In 2014, approximately 9.5% of the cattle slaughtered in the U.S. were culled dairy cows: cows that can no longer be seen as an economic asset to the dairy farm. These animals may be sold due to reproductive problems or common diseases of milk cows such as mastitis and lameness.

Calf

Most heifers (female calves) are kept on farm to be raised as a replacement heifer, a female that will be bred and enter the production cycle. Market calves are generally sold at two weeks of age and bull calves may fetch a premium over heifers due to their size, either current or potential. Calves may be sold for veal, or for one of several types of beef production, depending on available local crops and markets. Such bull calves may be castrated if turnout onto pastures is envisaged, in order to render the animals less aggressive. Purebred bulls from elite cows may be put into progeny testing schemes to find out whether they might become superior sires for breeding. Such animals may become extremely valuable.

Most dairy farms separate calves from their mothers within a day of birth to reduce transmission of disease and simplify management of milking cows. Studies have been done allowing calves to remain with their mothers for 1, 4, 7 or 14 days after birth. Cows whose calves were removed longer than one day after birth showed increased searching, sniffing and vocalizations. However, calves allowed to remain with their mothers for longer periods showed weight gains at three times the rate of early re-movals as well as more searching behavior and better social relationships with other calves.

After separation, some young dairy calves subsist on commercial milk replacer, a feed based on dried milk powder. Milk replacer is an economical alternative to feeding whole milk because it is cheaper, can be bought at varying fat and protein percentages, and is typically less contaminated than whole milk when handled properly. Some farms pasteurize and feed calves milk from the cows in the herd instead of using replacer. A day-old calf consumes around 5 liters of milk per day.

Cattle are social animals; their ancestors tended to live in matriarchal groups of mothers and offspring. The formation of "friendships" between two cows is common and long lasting. Traditionally individual housing systems were used in calf rearing, to reduce the risk of disease spread and provide specific care. However, due to their social behaviour the grouping of offspring may be better for the calves' overall welfare. Social interaction between the calves can have a positive effect on their growth. It has been seen that calves housed in grouped penning were found to eat more feed than those in single pens, suggesting social facilitation of feeding behaviour in the calves. Play behaviour in pre-weaned dairy calves has also been suggested to help build social skills for later in life. It has been seen that those reared in grouped housing are more likely to become the dominant cattle in a new combination of animals. These dominant animals have a priority choice of feed or lying areas and are generally stronger animals. Due to these reasons, it has become common practice to group or pair calves in their housing. It has become common within Canada to see paired or grouped housing in outdoor hutches or within an indoor pack penning.

Bull

A bull calf with high genetic potential may be reared for breeding purposes. It may be kept by a dairy farm as a herd bull, to provide natural breeding for the herd of cows. A bull may service up to 50 or 60 cows during a breeding season. Any more and the sperm count will decline, leading to cows "returning to service" (to be bred again). A herd bull may only stay for one season, as when most bulls reach over two years old their temperament becomes too unpredictable.

Bull calves intended for breeding commonly are bred on specialized dairy breeding farms, not production farms. These farms are the major source of stocks for artificial insemination.

Milk Production Levels

The dairy cow will produce large amounts of milk in its lifetime. Production levels peak at around 40 to 60 days after calving. Production declines steadily afterwards until milking is stopped at about 10 months. The cow is "dried off" for about sixty days before calving again. Within a 12 to 14-month inter-calving cycle, the milking period is about 305 days or 10 months long. Among many variables, certain breeds produce more milk than others within a range of around 6,800 to 17,000 kg (15,000 to 37,500 lbs) of milk per year.

The Holstein Friesian is the main breed of dairy cattle in Australia, and said to have the "world's highest" productivity, at 10000L of milk per year. The average for a single dairy cow in the US in 2007 was 9164.4 kg (20,204 lbs) per year, excluding milk consumed by her calves, whereas the same average value for a single cow in Israel was reported in the Philippine press to be 12,240 kg in 2009. High production cows are more difficult to breed at a two-year interval. Many farms take the view that 24 or even 36 month cycles are more appropriate for this type of cow.

Dairy Cows, Collins Center, New York

Dairy cows may continue to be economically productive for many lactation cycles. In theory a longevity of 10 lactations is possible. The chances of problems arising which may lead to a cow being culled are high, however; the average herd life of US Holstein is today fewer than 3 lactations. This requires more herd replacements to be reared or purchased. Over 90% of all cows are slaughtered for 4 main reasons:

- Infertility – failure to conceive and reduced milk production.

  Cows are at their most fertile between 60 and 80 days after calving. Cows remaining "open" (not with calf) after this period become increasingly difficult to breed, which may be due to poor health. Failure to expel the afterbirth from a previous pregnancy, luteal cysts, or metritis, an infection of the uterus, are common causes of infertility.

- Mastitis – a persistent and potentially fatal mammary gland infection, leading to high somatic cell counts and loss of production.

  Mastitis is recognized by a reddening and swelling of the infected quarter of the udder and the presence of whitish clots or pus in the milk. Treatment is possible with long-acting antibiotics but milk from such cows is not marketable until drug residues have left the cow's system, also called withdrawal period.

- Lameness – persistent foot infection or leg problems causing infertility and loss of production.

  High feed levels of highly digestible carbohydrate cause acidic conditions in the cow's rumen. This leads to Laminitis and subsequent lameness, leaving the cow vulnerable to other foot infections and problems which may be exacerbated by standing in faeces or water soaked areas.

- Production – some animals fail to produce economic levels of milk to justify their feed costs.

  Production below 12 to 15 litres of milk per day is not economically viable.

Cow longevity is strongly correlated with production levels. Lower production cows live longer than high production cows, but may be less profitable. Cows no longer wanted for milk production are sent to slaughter. Their meat is of relatively low value and is generally used for processed meat. Another factor affecting milk production is the stress the cow is faced with. Psychologists at the University of Leicester, UK, analyzed the musical preference of milk cows and found out that music actually influences the dairy cow's lactation. Calming music can improve milk yield, probably because it reduces stress and relaxes the cows in much the same way as it relaxes humans.

Cow Comfort and its Effects on Milk Production

Certain behaviors such as eating, rumination, and lying down can be related to the health of the cow and cow comfort. These behaviors can also be related to the productivity of the cows. Likewise, stress, disease, and discomfort will have a negative effect on the productivity of the dairy cows. Therefore, it can be said that it is in the best interest of the farmer to increase eating, rumination, and lying down and decrease stress, disease, and discomfort to achieve the maximum productivity possible. Also, estrous behaviors such as mounting can be a sign of cow comfort, since if a cow is lame, nutritionally deficient, or are housed in an over crowded barn, the performance of estrous behaviors will be altered.

Feeding behaviors are obviously important for the dairy cow, as feeding is how the cow will ingest dry matter, however, the cow must ruminate to fully digest the feed and utilize the nutrients in the feed. Dairy cows with good rumen health will likely be more profitable than cows with poor rumen health, as a healthy rumen will aid in the digestion of nutrients. An increase in the time a cow spends ruminating is associated with the increase in health and an increase in milk production. The productivity of dairy cattle is most efficient when the cattle have a full rumen . Also, the standing action while feeding after milking has been suggested to enhance udder health. The delivery of fresh feed while the cattle are away for milking stimulates the cattle to fed upon return, potentially reducing the prevalence of mastitis as the sphincters have time to close while standing This makes the pattern of feeding directly after being milked an ideal method of increasing the efficiency of the herd.

Cows have a high motivation to lie down so farmers should be conscious of this, not only because they have a high motivation to lie down, but also because lying down can increase milk yield. When the lactating dairy cow lies down, blood flow is increased to the mammary gland which in return results in a higher milk yield.

To ensure that the dairy cows lie down as much as needed, the stalls must be comfortable. Put very simply, a stall should have a rubber mat, bedding, and be large enough for the cow to lie down and get up comfortably. Signs that the stalls may not be comfortable enough for the cows are the cows are standing, either ruminating or not, instead of lying down, or perching, which is when the cow has its front end in the stall and their back end out of the stall.

There are 2 types of housing systems in dairy production, free style housing and tie stall. Free style housing is where the cow is free to walk around and interact with its environment and other members of the herd. Tie stall housing is when the cow is chained to a stantion stall with the milking units and feed coming to them.

By-products and Processing

Pasteurization is the process of heating milk to a high enough temperature for a short period of time to kill the microbes in the milk and increase keep time and decrease spoilage time. By killing the microbes, decrease the transmission of infection, and elimination of enzymes the quality of the milk and the shelf life increases. Pasteurization is either completed at 63 °C for 30 minutes or a flash pasteurization is completed for 15 seconds at 72 °C. By-products of milk include butterfat, cream, curds, and whey. Butterfat is the main lipid in milk. The cream contains 18–40% butterfat. The industry can be divided into 2 market territories; fluid milk and industrialized milk such as yogurt, cheeses, and ice cream.

Whey protein makes up about 20% of milk's protein composition and is separated for the casein (80% of milk's protein make up) during the process of curdling cheese. This protein is commonly used in protein bars, beverages and concentrated powder, due to its high quality amino acid profile. It contains levels of both essential amino acids as well as branched that are above those of soy, meat, and wheat. "Diafiltered" milk is a process of ultrafiltration of the fluid milk to separate lactose and water from the casein and whey proteins. This process allows for more efficiency in cheese making and gives the potential to produce low-carb dairy products.

Reproduction

Since the 1950s, artificial insemination (AI) is used at most dairy farms; these farms may keep no bull. Artificial insemination uses estrus synchronization to indicate when the cow is going through ovulation and is susceptible to fertilization. Advantages of using AI include its low cost and ease compared to maintaining a bull, ability to select from a large number of bulls, elimination of diseases in the dairy industry, improved genetics and improved animal welfare Rather than a large bull jumping on a smaller heifer or weaker cow, AI allows the farmer to complete the breeding procedure within 5 minutes with minimum stress placed on the individual female's body.

Dairy cattle are polyestrous, meaning they cycle continuously throughout the year. They tend to be on a 21 day estrus cycle, however for management purposes, some operations sync their cows or heifers using synthetic hormones in order to have the ideal time for breeding and later calving. These hormones are short term and are only used when necessary. For example, one common protocol for synchronization involves an injection of GnRH (gonadotrophin releasing hormone). which increases the levels of follicle stimulating hormone and luteinizing hormone in the body. Then, seven days later prostaglandin F2-alpha is injected, followed by another GnRH injection 48 hours later. This protocol causes the animal to ovulate 24 hours later.

Estrus is often called standing heat in cattle and refers to the time in their cycle where the female is receptive towards the male. Estrus behaviour can be detected by an experienced stockman. These behaviours can include standing to be mounted, mounting other cows, restlessness, decreased milk production, and decreased feed intake.

More recently, embryo transfer has been used to enable the multiplication of progeny from elite cows. Such cows are given hormone treatments to produce multiple embryos. These are then 'flushed' from the cow's uterus. 7–12 embryos are consequently removed from these donor cows and transferred into other cows who serve as surrogate mothers. The result will be between 3 and 6 calves instead of the normal single, or rarely, twins.

Hormone use

Hormone treatments are sometimes given to dairy cows in some countries to increase reproduction and to increase milk production.

The hormones are used to produce multiple embryos have to be administered at specific times to dairy cattle to induce ovulation. Frequently, for economic considerations, these drugs are also used to synchronise a group of cows to ovulate simultaneously. The hormones prostaglandin, gonadotropin-releasing hormone, and progesterone are used for this purpose and sold under the brand names Lutalyse, Cystorelin, Estrumate, Estroplan, Factrel, Prostamate, Fertagyl, Insynch, and Ovacyst. They may be administered by injection.

About 17% of dairy cows in the United States are injected with Bovine somatotropin, also called recombinant bovine somatotropin (rBST), recombinant bovine growth hormone (rBGH), or artificial growth hormone. The use of this hormone increases milk production from 11%–25%. The U.S. Food and Drug Administration (FDA) has ruled that rBST is harmless to people. The use of rBST is banned in Canada, parts of the European Union, as well as Australia and New Zealand.

In Canada, Canadian Dairy farmers have high screening procedures they have to go through every time the milk is retrieved from the farm; if the regulations are not met the milk does not get loaded onto the truck for further processing. There is to be no medication or hormones in the milk for safety reasons.

Nutrition

Dairy cattle at feeding time

Nutrition plays an important role in keeping cattle healthy and strong. Implementing an adequate nutrition program can also improve milk production and reproductive performance. Nutrient requirements may not be the same depending on the animal's age and stage of production.

Forages, which refer especially to hay or straw, are the most common type of feed used. Cereal grains, as the main contributors of starch to diets, are important in meeting the energy needs of dairy cattle. Barley is one example of grain that is extensively used around the world. Barley is grown in temperate to subarctic climates, and it is transported to those areas lacking the necessary amounts of grain. Although variations may occur, in general, barley is an excellent source of balanced amounts of protein, energy, and fiber.

Ensuring adequate body fat reserves is essential for cattle to produce milk and also to keep reproductive efficiency. However, if cattle get excessively fat or too thin, they run the risk of developing metabolic problems and may have problems with calving. Scientists have found that a variety of fat supplements can benefit conception rates of lactating dairy cows. Some of these different fats include oleic acids, found in canola oil, animal tallow, and yellow grease; palmitic acid found in granular fats and dry fats; and linolenic acids which are found in cottonseed, safflower, sunflower, and soybean. It is also important to note that proper levels of fat also improve cattle longevity.

Using by-products is one way of reducing the normally high feed costs. However, lack of knowledge of their nutritional and economic value limits their use. Although the reduction of costs may be significant, they have to be used carefully because animal may have negative reactions to radical changes in feeds, (e.g. fog fever). Such a change must then be made slowly and with the proper follow up.

Pesticide use

A survey of the primary dairy producing areas in the US indicated that 13 percent of lactating animals were treated with insecticides permethrin, pyrethrin, coumaphos, and dichlorvos primarily by daily or every-other-day coat sprays. Workers, particularly in stanchion barns, may be exposed to higher than recommended amounts of these pesticides.

Breeds

According to the Purebred Dairy Cattle Association, PDCA, there are 7 major dairy breeds in the United States. These are: Holstein, Brown Swiss, Guernsey, Ayrshire, Jersey, Red and White, and Milking Shorthorn.

Holstein cows either have distinct white and black markings, or distinct red and white markings. Holstein cows are the biggest of all U.S. dairy breeds. A full mature Holstein cow usually weighs around 1,500 pounds (700 kg) and is 58 inches (147 cm) tall at the shoulder. They are known for their outstanding milk production among the main breeds of dairy cattle. An average Holstein cow produces around 23,000 pounds (10,000 kg) of milk each lactation. Of the 9 million dairy cows in the U.S., approximately 90% of them are of the Holstein descent. The top breed of dairy cow within Canada's national herd category is Holstein, taking up 93% of the dairy cow population, have a production rate of 10,257 kilograms (22,613 lb) of milk per cow that contains 3.9% butter fat and 3.2% protein.

Brown Swiss cows are widely accepted as the oldest dairy cattle breed, originally coming from a part of northeastern Switzerland. Some experts think that the modern Brown Swiss skeleton is similar to one found that looks to be from around the year 4000 B.C. Also, there is evidence that monks started breeding these cows about 1000 years ago.

The Ayrshire breed first originated in the County of Ayr in Scotland. It became regarded as a well established breed in 1812. The different breeds that were crossed to form the Ayrshire are not exactly known. However, there is evidence that several breeds were crossed with the native cattle to create the breed.

Guernsey cows originated just off the coast of France on the small Isle of Guernsey. The breed was first known as a separate breed around 1700. Guernseys are known for their ability to produce very high quality milk from grass. Also, the term "Golden Guernsey" is very common as Guernsey cattle produce rich, yellow milk rather than the standard white milk other cow breeds produce.

The Jersey breed of dairy cow originated on a small island located off the coast of France called Jersey. Despite being one of the oldest breeds of dairy cattle they now only occupy 4% of the Canadian National Herd. Purebred Jersey cows, according to available data, have been in the UK area since about the year 1741. When they were first bred in this area, they were not known as Jerseys, but rather as the related Alderneys. The period between 1860 and around 1914 was a popular time for Jerseys. In this time span, many countries other than the United States started importing this breed, including Canada, South Africa, and New Zealand, among others.

Among the smallest of the dairy breeds, the average Jersey cow matures at approximately 900 pounds (410 kg), with a typical weight range between 800 and 1,200 pounds (360–540 kg). According to North Dakota State University, the fat content of the Jersey cow's milk is 4.9 percent. It is also the highest in protein, at 3.8 percent. This high fat content means the milk is often used for making ice cream and cheeses. According to the American Jersey Cattle Association, Jerseys are found on 20 percent of all US dairy farms and are the primary breed in about 4 percent of dairies.

Amongst the Bos indicus, the most popular dairy breed in the world is Sahiwal of the Indian subcontinent. It does not give as much milk as the Taurine breeds, but it is by far the most suitable breed for warmer climates. Australian Friesian Sahiwal and Australian Milking Zebu have been developed in Australia using Sahiwal genetics. Gir, another of the Bos Indicus breeds, has been improved in Brazil for its milk production and is widely used there for dairy.

Animal Welfare

Animal welfare refers to both the physical and mental state of an animal, and how it is coping with its situation. An animal is considered in a good state of welfare if it is able to express its innate behaviour, comfortable, healthy, safe, well nourished, and is not suffering from negative states such as distress, fear and pain. Good animal welfare requires disease prevention and veterinary treatment, appropriate shelter, management, nutrition, and humane handling. If the animal is slaughtered then it is no longer "good animal welfare". It is the human responsibility of the animals' wellbeing in all husbandry and management practices including humane euthanasia. Welfare differs greatly compared to animal rights. Welfare approves animal use by humans, wheres animal rights do not believe in the use of animals in any circumstance including companion animals.

Proper animal handling, or stockmanship, is crucial to dairy animals' welfare as well as the safety of their handlers. Improper handling techniques can stress cattle leading to impaired production and health, such as increased slipping injuries. Additionally, the majority of nonfatal worker injuries on a dairy farm are from interactions with cattle. Dairy animals are handled on a daily basis for a wide variety of purposes including health-related management practices and movement from freestalls to the milking parlor. Due to the prevalence of human-animal interactions on dairy farms, researchers, veterinarians, and farmers alike have focused on furthering our understanding of stockmanship and educating agriculture workers. Stockmanship is a complex concept that involves the timing, positioning, speed, direction of movement, and sounds and touch of the handler.

A recent survey of Minnesota dairy farms revealed that 42.6% of workers learned stockmanship techniques from a family members, and 29.9% had participated in stockmanship training. However, as the growing U.S. dairy industry increasingly relies on an immigrant workforce, stockmanship training and education resources will become more pertinent. Clearly communicating and managing a large culturally diverse workforce brings new challenges such as language barriers and time limitations. Organizations like the Upper Midwest Agriculture Safety and Health Center (UMASH) offer resources such as bilingual training videos, fact sheets, and informational posters for dairy worker training. Additionally the Beef Quality Assurance Program offer seminars, live demonstrations, and online resources for stockmanship training.

In order for the cows to reach high performance in milk yields and reproduction, they must be in great condition and comfortable in the system of which they are reared. Once an individual's welfare is reduced, so does her efficiency and production. This creates more cost and time on the operation, therefore most farmers strive to create a healthy, hygienic, atmosphere for their cattle. As well as provide quality nutrition that will keep the cows yield high.

The practice of dairy production in a factory farm environment has been criticized by animal rights activists. Some of the ethical reasons regarding dairy production cited include how often the dairy cattle are impregnated, the separation of calves from their mothers, and the fact that the cows are considered "spent" and culled at a relatively young age, as well as environmental concerns regarding dairy production.

The production of milk requires that the cow be in lactation, which is a result of the cow having given birth to a calf. The cycle of insemination, pregnancy, parturition, and lactation is followed by a "dry" period of about two months before calving, which allows udder tissue to regenerate. A dry period that falls outside this time frames can result in decreased milk production in subsequent lactation. Dairy operations therefore include both the production of milk and the production of calves. Bull calves are either castrated and raised as steers for beef production or veal.

Animal rights groups such as Mercy for Animals also raise welfare concerns by citing undercover footage showing abusive practices at factory farms.

Beef Cattle Farming

Beef production is a large and important segment of South African farming. Beef farming works well with other agricultural enterprises like grain (in particular), orchard, vegetable, or other crop operations. Cattle can make efficient use of feed resources that have little alternative use, such as crop residues, marginal cropland, and land not suitable for tillage, or land that cannot produce crops other than grass.

For people who own land but work full-time off the farm, a beef enterprise can be the least labour-intensive way to utilize their land. A cattle enterprise can use family or surplus labour. Calving, weaning, vaccinations, castration, and weighing can be planned for times when labour is available.

Consider your resources, the land available, and your level of interest and capabilities before deciding to engage in the cattle business. Identify why you want to raise cattle and set goals to achieve the most constant economic return or personal satisfaction. Your goals must be clearly defined, firmly fixed, achievable, and have a realistic time frame. Otherwise, your operation will lack focus.

Production Options

There are few types of small-scale cattle enterprises.

- *Growing and feeding systems*--In these operations, calves or wearners are either raised or purchased and then are fed (fattened for slaughter). Included in this category are operations specializing in producing cattle for home use.

- *Breeding herds*--A breeding herd consists of cows and bulls that are used to produce calves for sale as breeding or feeder animals.

- *Combinations of growing, feeding, and breeding herds.*

Success of your operation will entirely depend on adapting a strategy that fits your needs and capabilities.

Growing and Feeding Operations

In a weaners (yearlings) *operation,* weaner calves are acquires after weaning at 10 to

15 months of age. They can be fed out and marketed in less than a year from the time of purchase. Thus, the investment on each calf is returned within a comparatively short time. This type of operation may not require much land, but adequate facilities are essential so that animals can be kept comfortable and under control.

Some good enterprises are based pasture operations. Weaned calves are purchased in early spring, go on pasture (when the grass is at its best with regard to productivity), and are sold when the pasture season is over. On the other hand, calves cost less during winter; therefore, depending on the cost of winter feed, this may be the best time to purchase cattle for the next pasture season. Purchase price and selling price greatly influence profitability in this enterprise.

Managing Calves

Keep calves in an area that allows you to observe them quite closely for two weeks. This enables you to prevent the spread of disease. Calves should have access to plenty and fresh water and feed. Working the calves requires a lot of patience, as they are easily excited and stressed. Consult a veterinarian for a health program that lowers the risk of disease for newly received calves.

Breeding Herds

Establishing a breeding herd is a long-term objective. It also requires more land than in a situation where weaner feeding program is implemented. Consider how your available resources match your long-term objectives. There must be adequate feed, water, and fences to accommodate a year-round operation.

Decide whether to have registered pure-bred cattle or commercial cattle. Income from a commercial beef herd comes mainly from the sale of calves and old or cull animals. Sale of breeding stock is the main source of income from registered cattle. Care and management of registered cattle is more intensive than for commercial cattle.

Developing a Registered Herd

If your objective is to raise registered cattle and supply breeding animals to other cattle producers, it may be necessary to make large capital investments in purebred stock. Development of a registered herd means that both the sire and dam must be purebred and registered with the same national breed association on the stud book. You must keep accurate records and register the desirable purebred calves to be retained for breeding stock.

If you raise bulls for the beef industry, you must develop a selection program based on characteristics of economic importance, such as fertility, mothering ability, ease of calving, growth rate, and carcass merit. Also, use great care in the selection of breeding females, as considerable time and expense are involved. Competition is keen with already-established herds. However, there are successful registered herds with only 20 to 40 cattle.

Developing a Commercial Herd

The criteria for selection, or selling points, of good commercial cows depend on size, quality, age, condition, stage of pregnancy, and market price. You should select breed and cow size to match your feed resources and topography. Local extension officers can give you an idea of what breeds are best suited to your area. Crossbreeding (mating animals from two or more breeds) can be an advantage in a commercial cow herd. Capitalizing on the merits of several breeds, plus the extra vigor from crossbred calves, may give you a competitive edge in the market. Remember that advances in genetic merit probably will not be realized for several years.

Purchasing Cattle

There are many sources of good cattle, both registered (stud breeds) and commercial. Usually it's best to purchase from a successful and reputable breeder. They usually sell only sound cattle as breeding animals and they are helpful in giving advice to less experienced producers.

If you are inexperienced, it might be best to buy good, young, bred cows that have calved at least once. This reduces problems associated with calving heifers. If you purchase open heifers, you should breed them to a bull that has the traits for easy calving.

Managing a Cow-calf Herd

The major concern of cattle producers is profit. For a cow-calf herd, profits are determined by the percent calf crop (the number of calves weaned per cows bred), the weaning weight of the calves, the costs of maintaining breeding animals, and, ultimately, the sale price of the calves. Because your entire program depends on the fitness of the breeding animals, it is essential to maintain good herd health by not allowing the cattle to become too fat or too thin. Cows do not milk as well and may have problems calving or getting bred if they are overweight or underweight (refer to Body Score Conditioning in beef cattle). Bulls that are not in good condition may perform poorly during the breeding season.

Breeding Season

It is ideal to have a controlled breeding season, rather than allowing the bull to run with the cows continuously. A month to one and half or even two months breeding season is recommended. The resulting shortened calving season increases the possibility of having a uniform set of calves to sell at market time. Cattle of similar breeding and size usually bring more money. Another advantage is that you can concentrate your work with cows during calving into a short span, instead of having it strung out for months.

Cattle have a 283 days gestation period. Select breeding dates so that cows will calve at the time of year you desire. Considerations in determining calving season include weather conditions and the ability to match feed resources with the cows' requirements.

A quality sire is essential to maintain a good, healthy herd. The rule of thumb in terms of bull to cow ratio is 1 bull to 25 cows. The ratio varies depending on the bull's age and health, and the size of pasture. Small herd owners have the following options for obtaining a good-quality bull:

- You can buy a bull in cooperation with another farmer.

- You can lease or borrow a sire from a neighbour.

However, using a bull increases the risk of diseases. Bulls also may pose a safety risk, so treat them with respect.

Another good breeding option is artificial insemination (AI). If you use this method, you should synchronize oestrus in the herd. This process may require the aid of a veterinarian.

The last consideration of the breeding season is pregnancy testing the cows. The test helps determine which cows should be culled from the herd to avoid the costs of wintering a cow that is not pregnant. Veterinarians offer pregnancy testing services.

Calving

This aspect of beef cattle management requires experience and skill. If you are inexperienced, it is recommended that you contact your veterinarian and/or extension officers for advice on calving management.

One of the simplest ways to add to the value of your calves is to make sure they are well fed, properly castrated, dehorned, vaccinated, and clearly identified.

The most important thing to remember when working calves is to stress them as little as possible. You can learn how to castrate, dehorn, and give vaccinations under the supervision of an experienced cattle producer or veterinarian.

Keeping Performance Records

Keeping records enables you to cull poor performers and maintain good overall herd healthy and vigour. Examples of helpful calf records include birth weight, weaning weight, and average daily gain.

Combinations of Breeding, Growing, and Feeding

Most calves produced in small commercial herds are marketed as weaned calves weighing from 250 to 300 kg. Other options include the following:

- Wean the calves, winter them, and sell them as yearlings.

- Creep feed calves while the animals are still nursing, put them on full feed after weaning, and then sell them as slaughter cattle at 12 to 16 months of age.

- Wean calves, winter them on a growing ration, then graze them during spring and early summer and finish them to slaughter weight at 18 to 24 months of age.

Facilities and Equipments

Producing beef cattle on a small farm does not require elaborate or expensive housing or facilities. Under wide range of weather conditions, cattle do very well outside. One method is to allow animals to have access to an open-air pole shelter. In an enclosed building, proper ventilation is important to maintain good health.

Design facilities to make your job easy and safe and to minimize your expenditure of time and labor. An effective working facility consists of a crush pens, a head clamp and a squeeze chute.

The crush pen is needed for vaccinations, deworming, etc. The neck clamp is needed if you must aid a cow with calving. The pens and narrow alley help confine animals that need to be handled and driven into the crush pen or neck clamp.

Well-designed handling facilities help to minimize animal confusion and stress. Poorly designed facilities increase stress on the animals and may cause poor performance, which can affect meat quality. Use of electric prods is not recommended because they cause animals unnecessary pain and stress.

It is important to maintain the quality of feed. Store forages (including hay, straw, or silage) and grains in a dry building free from rodents. Forages lose nutritional value when exposed to direct sunlight. Wet hay loses feed value and palatability and presents a safety hazard due to combustion and development of moulds. Rodents can damage feed and spread disease.

Feeders reduce waste and prevent the spread of many internal parasites and other cattle diseases. You can buy many kinds of manufactured feeders. Or, you can build them out of materials on hand.

An adequate, year-round supply of clean, fresh water is basic to any successful cattle farming. Many types of water troughs are available from local feed or farm supply stores. You can recycle old barrels and bathtubs to make functional troughs; be sure to clean them thoroughly prior to use.

Pens, feedlots, and crush pen should be located at a convenient distance from feed storage facilities. These areas should be well drained, with drainage moving away from feed storage, working facilities, and roads. It is important to make these areas accessible to tractors for easy feeding and cleaning.

Proper transportation is a must for your cattle. A 1-ton or 3/4-ton truck and trailer are convenient for any beef operation. A truck also is useful for transporting and dispersing hay.

Feeding Beef Cattle

Unlike humans, cattle have a ruminant digestive system. Their stomachs are made up of four parts. Ruminant microorganisms in the first three parts enable cattle to digest fibrous feeds that single stomach animals cannot. This microbial breakdown produces essential nutrients such as amino acids and B vitamins. The presence of these nutrients makes beef very useful for human consumption.

Nutritional Needs

Cattle require protein, energy, water, fat, minerals, and vitamins. The amounts vary according to environment, the cow's age, time of year, and production goals and stages. Availability of feedstuffs also varies by location and season. Up to 75% of the cost of raising an animal goes to feed under intensive feeding system.

Protein and carbohydrate levels adequate for growth and maintenance normally are found in high-quality legume hay, such as Lucerne and clover. Poor-quality feeds, such as cereal straw, grass straws, or rain-damaged hay, require protein or energy supplements. You can purchase supplements from your feed supplier.

Beef cattle normally do not need vitamin A, B, or E supplementation. They can get these vitamins from normal-quality feedstuffs. However, a vitamin A deficiency can result from feeding dry, bleached-out hay. Symptoms of vitamin A deficiency include watery eyes, rough hair coat, night blindness, and poor gains.

Vitamin D is formed by the action of sunlight on animal tissues. If you confine your cattle to a barn or stall for extended periods of time, vitamin D deficiency may become a problem.

Minerals are inorganic compounds that contribute to bones, teeth, protein, and lipid functions of the body. Minerals are provided through natural feeds and supplementation.

There are three main categories of mineral supplements:

- *Salt,* which usually is sold as iodized salt and does not contain other Minerals

- *Trace mineralized salt,* which consists of a large percentage of salt and traces of some or all of the following: copper, iron, iodine, cobalt, manganese, selenium, and zinc

- *Mineral mixes,* which usually contain major minerals such as calcium and phosphorus as well as trace minerals and some salt

You can provide supplements as licks or mix them into feed. The composition of needed salt or mineral supplements varies depending on your locale and feedstuffs. Clean water is essential and must be provided at all times. Under normal conditions, cattle consume 20 to 70 litters of water per day depending on size, age, and weather. Heat dramatically increases water consumption.

Types of Feed

Feedstuffs are categorized as concentrates or roughages. Concentrates are high in digestible nutrients. Grains and protein supplements are examples of concentrates. Roughages are feedstuffs that are low in digestible nutrients. Examples of roughages include hay, pasture, and silage.

The percentage of roughage and concentrate in beef cattle rations depends on the type of animal being fed. For example, feedlot steers are fed mostly grain and a little roughage, while bred cows may be wintered on good-quality roughage alone. *Caution:* High-quality legume hay such as Lucerne may cause bloat (most probably if grazed by hungry animals). As a general rule, beef cattle consume up to 3 kg of feed per day for each 100 kg of body weight. A 300 kg weaned calf, for example, will eat 9 kg of high-quality Lucerne hay per day.

Cattle usually weigh 250 to 300 kg before they are placed on a high-grain (highenergy) ration. This diet is fed until slaughter weight is achieved.

If you feed out cattle for slaughter, you can purchase feed or grow and mix it at home. If only a few animals are being finished, it may be more economical to purchase the mixed ration from a feed dealer.

Growth promotants, including implants, may have a place in your operation. They are used widely in the industry and have been proven safe. Ionophores are feed additives that decrease rumen upset, increase feed efficiency, and increase daily gains. These chemicals can improve gain significantly; however, they do not compensate for poor management.

Health Problems

Cattle of all ages, particularly young, growing cattle, are subject to a variety of ailments. They range from mild conditions to severe infectious diseases that may cause death within 24 hours. The cost of caring for sick cattle can seriously reduce your profit margin. With the increasing need to cut production costs, good herd health care is very important for any beef operation. Prevention is the easiest and cheapest method of disease control. Clean sheds, lots, and feed and water troughs give disease less chance to get started. A sound vaccination program, parasite control, and frequent observation of the herd also help to reduce the occurrence of illness.

You can recognize a sick animal first by its abnormal behavior or physical appearance. Droopy ears, loss of appetite, head down, scouring (diarrhea), or inactivity may indicate illness. A high temperature usually indicates disease.

The best course of action is to find a sick animal quickly, treat it, and then work to eliminate the cause of the sickness. If one or two animals come down with a disease, the rest of the herd has been or will be exposed to it. Health problems are more common during and after periods of stress, including calving, weaning, shipping, working or moving the cattle, and extreme weather conditions. Stress can reduce an animal's ability to resist infectious agents. After a period of stress, give extra attention to your animals' health.

Common Cattle Diseases

The following are five of the more common health problems that beef producers encounter. You also need to be aware of other diseases that affect the health of livestock in your region.

Respiratory Diseases

Respiratory diseases are common in cattle. A number of factors contribute to an outbreak: inadequate nutrition, stress, and viral or bacterial infection. Good management and vaccination of cows and calves is the best way to prevent outbreaks of respiratory disease. Your veterinarian can help you develop a program to reduce losses on your ranch and in the feedlot.

Brucellosis

Brucellosis is a serious disease. It causes abortion and sterility in cattle. Under South African Animals Health Acts, they effectively outline brucellosis as notifiable disease. Vaccination is required for all heifers. Brucellosis most commonly enters a herd through the purchase of infected cattle. To help prevent brucellosis from entering your herd, vaccinate all heifers between ages 4 to 10 months, and purchase only brucellosis-vaccinated cattle.

External Parasites

External parasites include horn flies, face flies, stable flies, ticks and lice. The largest health problem comes from the additional stress these insects cause to animals. When infested, cattle spend more time in the shade and don't graze, which causes poor performance. You can reduce these problems by using flyrepellent ear tags or another parasite control treatment. Eliminating the areas where pests reproduce also helps to reduce the severity of external parasites. Pour-on and dips are effecting in treating animals infested by tick.

Internal Parasites

Internal parasites such as roundworms, lungworms, and liver flukes commonly occur in cattle. These hidden parasites cause poor performance and occasionally kill young animals. Cattle are likely to pick up internal parasites when they graze established pastures. Internal parasites also can be a big problem in confined areas.

Invasion of the stomach or intestinal wall by a parasite leads to poor digestion of nutrients and damage to organs. Signs of parasite infestation include scouring, rough hair coat, poor gains, and potbelly appearance.

Use dewormers at strategic times during the year to reduce the numbers of internal parasites. Use fecal sampling to determine the severity of the infestation and the type of dewormer that will be effective.

Disease Control

Vaccinations and parasite controls are available for many of the diseases affecting cattle. The choice of remedy and time of application depend on a variety of things, including the animal's nutritional level, disease prevalence in the herd, and the region in which the cattle are located. Local veterinarian should be consulted for a vaccination program according to the conditions existing at that area.

Marketing

It may not be easy to determine how and where to market your animals. The choice of market outlet depends on the class and grade of the cattle. Thus, the method of marketing usually is different for fed cattle, feeder, or purebred cattle.

There are many different methods of marketing cattle, but most livestock are marketed through one of three channels: auction, carcass grade and informal market (ceremonies, funerals and rituals) basis. The informal and auction markets are for both fed (feedlots) and feeder (farmers) cattle, while the carcass grade and weight basis is primarily for fed cattle.

Informal Marketing

Direct selling, or farm selling, refers to sales of livestock directly to packers, local dealers, or farmers without the use of agents or brokers. The sale usually takes place on the farm, or some other non-market buying station or collection yard. This method does not involve a recognized market. Sellers who participate in direct-market should be aware of possible regulations regarding Animal Welfare.

Niche Marketing

A producer often can develop a local or regional market for certain cuts of beef or specialty beef products. If this interests you, check into meat handling requirements, inspections, and permits that may be necessary. This type of marketing usually takes time to develop and also may require a consistent seasonal or yearly supply.

Auction Marketing

Livestock auctions or sales barns are trading centers where animals are sold by public bidding to the buyer who offers the highest price per hundredweight or per head. Auctions may be owned by individuals, partnerships, corporations, or cooperative associations.

Grades of Carcass Beef

Carcass beef sold to wholesale and retail outlets usually is graded to determine the quality and price. There are two categories of grades for beef: yield grade and quality grade in South Africa.

The system is especially designed to make the purchase of red meat simple for customers. The main characteristics used to classify beef carcass are the age of the animal and the fatness of the carcass.

The age of an animal is determined by the number of permanent incisor teeth; the more permanent incisors, the older the animal. The age of an animal is an indication of the tenderness of the meat - the meat of younger animal is more tender than that of older animal. The age classes are known as

A = meaning the youngest animals;

AB = meaning older animals;

B = meaning even older animals; and

C = meaning the oldest animals.

The fatness classes are known as class zero (no fat) to class 6 (excessively over fat). The roller mark on a carcass includes the age class (AAA, ABAB, BBB or CCC) and the fatness class (000, 111, 222, 333, 444, 555 or 666). When referring to a class of carcass, both the age class and fatness class are said, written, read or supposed to be listened to e.g. A1, AB2, C3 etc.

Each abattoir has a specific identity code which also appears in the roller mark. Consumers can therefore read in a roller mark on a carcass its class as related to carcass and eating quality characteristics and also be assured that the carcass originated from an approved abattoir and has passed a health inspection.

Budgets and Financial Records

Standard ranch records cover all production and financial management aspects of a beef operation. Use records to evaluate your business in terms of efficient use of resources and productivity. Records are important for ranch planning, tax reporting, and applying for credit.

Budgets

Decisions are only as good as the information on which they are based. Budgets provide the information for making ranch management decisions and are constructed to estimate the outcomes of future activities. Budgeting allows you to anticipate problems that you may encounter, and to alleviate or avoid them.

Financial Records

The way ranchers keep financial records varies, but the key is to use a system that provides the information you need to meet your responsibilities. The minimum set of financial records should include a balance sheet, a statement of cash flow, and an income statement.

There are several ways to keep accurate records. Hand-kept records are inexpensive and easy to store. On the other hand, this method may be slow and subject to errors. Retrieving information may be time consuming if extensive records are kept.

Computerized record systems are available, from simple checkbook balancing systems to sophisticated, double-entry accrual programs. Computerized systems for production records also are available in a range of features and reporting capabilities. Advantages include easy retrieval of information and reduced chance of mathematical errors. However, entering information takes time, and entries must be posted properly. If you choose a computer system, it should meet the requirements and objectives of your individual operation.

Ranch

Ranch, a farm, usually large, devoted to the breeding and raising of cattle, sheep, or horses on rangeland. Ranch farming, or ranching, originated in the imposition of European livestock-farming techniques onto the vast open grasslands of the New World.

Spanish settlers introduced cattle and horses into the Argentine and Uruguayan pampas and the ranges of Mexico early in the colonial period, and the herding of these animals spread readily into what is now the southwestern United States.

By the early 19th century the ranch had become an economic mainstay of the North American ranges. Its importance in the territorial United States was augmented as the progressive clearing and cultivation of grazing lands in the East drove cowherders west in pursuit of new pasture. The cowboy (q.v.) emerged during this period as essentially a rancher on horseback, who moved from camp to camp, grazing cattle on unfenced public ranges. Biannual roundups were held for branding calves and separating steers to be driven north and east for fattening and slaughter.

On the pampas of South America, where cattle and horses roamed freely for more than a century, the cowboy's southern counterpart, the gaucho (q.v.), first hunted huge semiwild herds independently and later worked for landowners, as the fenced estancia (estate) changed the face of the pampas.

The Homestead Act of 1862 in the United States generated the establishment of many grassland farms that were to expand into the huge western ranches of the late 19th century. Itinerant ranching reached its peak in the 1880s, when millions of cattle grazed the pastoral empire of the plains. Overstocking of ranges, the exceptionally hard winter of 1886–87, the passage of quarantine laws, increased railroad competition, and the encroachment of barbed-wire fencing all acted to check the northern cattle drives and diminish the glory of cattle country.

By the second quarter of the 20th century, nearly all livestock farming in the United States was sedentary. Huge ranches continued to exist, however, and, despite periods of fragmentation, the future of such enterprises seemed secure in the late 20th-century era of corporate agriculture. Open-range ranching has remained an important economic activity in Australia and New Zealand and in parts of Africa, where it was introduced in the late 19th century.

Ranch Occupations

The person who owns and manages the operation of a ranch is usually called a *rancher*, but the terms *cattleman, stockgrower,* or *stockman* are also sometimes used. If this individual in charge of overall management is an employee of the actual owner, the term *foreman* or *ranch foreman* is used. A rancher who primarily raises young stock sometimes is called a *cow-calf operator* or a *cow-calf man*. This person is usually the owner, though in some cases, particularly where there is absentee ownership, it is the ranch manager or ranch foreman.

The people who are employees of the rancher and involved in handling livestock are called a number of terms, including *cowhand, ranch hand,* and *cowboy.* People exclusively involved with handling horses are sometimes called wranglers.

Origins of Ranching

Ranching and the cowboy tradition originated in Spain, out of the necessity to handle large herds of grazing animals on dry land from horseback. During the Reconquista, members of the Spanish nobility and various military orders received large land grants that the Kingdom of Castile had conquered from the Moors. These landowners were to defend the lands put into their control and could use them for earning revenue. In the process it was found that open-range breeding of sheep and cattle (under the Mesta system) was the most suitable use for vast tracts, particularly in the parts of Spain now known as Castilla-La Mancha, Extremadura and Andalusia.

Automatic Milking

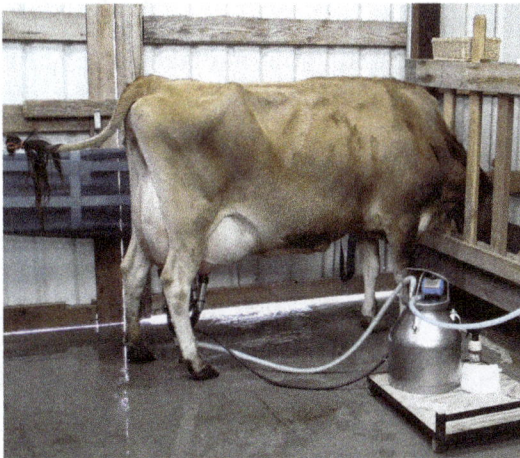

Robotic milking is an important link in the food chain defined by Lely as "from grass to glass". Managing a farm with milking robots requires a different approach compared to conventional milking. As market leader in fully automated milking, Lely has years of practical experience and research results that enable the company to give an accurate management advice for successful robotic milking.

Management, Milking

The milking robot supplies cow-related information unobtainable in a conventional situation, thereby making it possible to manage animals at an individual level in today's situation. Management by exception is the new challenge. The idea is that management should spend its valuable time concentrating on the cows requiring attention. Furthermore, the concept of the 'Licence to produce' is introduced focussing on sustainable dairy farming with respect to people, planet and profit. With the milking robot, it is possible to supply an individual cow with all her needs for an optimal health, production and well-being, without the extra labor. The basis of successful dairy farming is the healthy and happy individual cow.

Start up Procedure

Before the transition from conventional to robotic milking, it is very important to thoroughly consider what this will mean for your management. The robot will take a central place in the new situation and management should be adapted to this.

Preparation

In the preparation phase it is recommended to visit several farms of the same size and with the same type of barn, in order to get a good picture of the way robotic milking works and what it involves in terms of farm management. The experience of fellow dairy farmers is an important help in introducing the milking robot successfully into the farming processes. A well thought out (written) strategy is essential in the entire preparation phase: it will consist of a housing plan, a detailed plan on daily, weekly and monthly work routines, cow routines, and cow and farmer routings. The entire project should fit into a long-term strategy, so every step should be thoroughly evaluated: where will you treat cows, dry off cows, etc.

In this phase pay also attention to the following aspects:

- Formulate and write down goals for introducing the robots, to look back at them in the months after the introduction.

- Consider the feeding strategy. In many cases the crops and pellets need to be prepared to be fed in the robot, especially in places where till now only Total Mixed Ration (TMR) is used.

- Suitability for cows to be milked in the robot: singeing udders, cutting long hairs from the tail, optimizing claw health etc.

- Breed cows in such a way that you get less crossed teats or outstanding front teats.

- Prepare and understand the management program on-time, to prevent stress.

Barn Layout

The location of the Astronaut milking robots must be carefully planned for appropriate cow routing in the barn. The robot room must always have a clean entrance. The robots should be clearly visible and easily accessible for all cows. This means plenty of space around the robot and a clear, straight routing to and from the robots. It is important, particularly in a barn with more robots, that incoming and outgoing cows do not cross one another's path. More information on barn construction can be found in the brochure 'Barn design for robotic milking' or contact your local Lely Center.

Start Up

When starting with robotic milking, it is recommended to start with 50-60 cows per robot and divide the group into two subgroups. For the first three days, the cows are enticed into the robot three times a day. This should be done in a very calm and patient manner, to prevent the cows from having a negative association with the robot. Within these three days, 75% of the cows will go to the robots on their own, after which the routing gates can be removed. Set the cows free (free cow traffic) and collect the cows that have a milking interval of more than 10 hours, four times a day. The number of times the cows are fetched is gradually reduced to twice a day, fetching only the cows with a milking interval of more than 12 hours (or more than 10 kg (22 lbs) of milk).

This start-up procedure decreases waiting times and assures proper intake of dry matter (DM) and water. Collecting too many cows too soon results in low ranking cows waiting until they are collected. These cows will consider the robot as a crowded and dangerous place. Hence, they will wait until they are collected by the farmer. This stresses the importance of remaining calm and patient the first weeks and to follow the instructions on collecting cows.

Lely recommends for the first few days a minimum of two people available per robot to guide the cows and to control the X-link. After two or three days, it is sufficient to have one person per robot.

Organizing the Time Schedule

The farmer's daily time schedule changes when the robots are in use on the farm. The farmer no longer needs to milk the cows two or three times per day. This changes the routines that have become fixed in the daily business of farming. The milking robot gives farmers the opportunity to observe the cows in their own environment. Monitoring is simplified and any abnormal behavior of animals is easily noticed.

Changes in the time scheduling:

- Different, flexible working times because fixed milking times are history.
- More flexible working schedule.
- Shorter working times thanks to efficient management.
- Peaks in daily tasks are easily handled due to the freedom gained from milking by the robot.
- Time gained can be used outside the farm and/or for the management of individual animals.

Free Cow Traffic

With free cow traffic, the cows are free to the cubicles and water troughs without hindrance from fencing or selection gates. Experiences and observations of many farms all over the world show that free cow traffic is the basis for successful robotic milking. It increases profitability through an optimal production and healthy cows. Lely research on various forms of cow traffic shows that free cow traffic is characterized by a higher milk production with less labor and a reduced risk of mastitis. Farmers applying free cow traffic provide their cows with the five freedoms and by doing so they will get the most out of their herd.

Ten reasons to opt for free cow traffic:

1. More milk per cow (more rest and higher feed intake)
2. Less lameness (more rest)
3. Better for low ranking cows (less stress)
4. Better fat-to-protein ratio (higher roughage intake)
5. Higher feed efficiency and healthier rumens (through more frequent feed intake)
6. More freedom and improved animal well-being

7. Less labor and more milk per robot

8. Less mastitis (through less stress and more frequent milking)

9. Better social life of the farmer

10. Lower costs (investment in gates), higher profit

Visit Behavior

For good visit behavior of the cows, the robot has to be easily accessible. Firstly, this means that there should be sufficient free time on the robot (at least 10%) so a cow can enter the robot whenever she wants. Free time means the time that the robot is freely accessible (the door is open). When there is less free time, especially low-ranking timid animals will not be milked enough, simply because they do not have the opportunity or they are afraid to do so. Animals that are not milked at least two times/day have an increased risk of udder health problems.

Secondly, space in front and around the robot is critical to achieve sufficient visits to the robot. This area is the busiest part of the barn, so any obstructions will disrupt cow traffic and reduce visits to the robot. On a farm with 120 cows on two robots with 3 milkings and 1 refusal per day, this means 120x4 = 480 cow passages in front of the robots. For a good accessibility, the robot has to be visible and easy to reach from anywhere in the barn.

Milk Separator

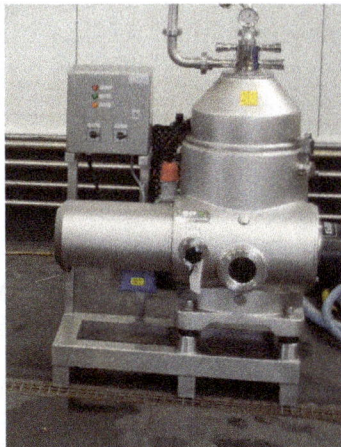

REDA centrifuges of RE-T serie are specifically designed to skim, clean and standardize milk, cream and whey.

These separators are specially built for dairy applications and match very high bowl speeds for the best performances in separation of solid particles with very small diameters (fat, dirty, spores and bacterias, etc). Standard design of these separators provides the product feeding under pressure by a soft inflow system. Then the product is accellerated to separation speed without damage being caused to any of the fat globules. The wide working surface, the high spinning speed and the extremely fast sludge expulsion lead to maximum separation and cleaning efficiency. The skimmed product and cream exit under pressure, therefore no recovery pumps are necessary. The range of process capacity goes from minimum 1500 l/h to maximum 50,000 l/h.

Mechanism

Manual rotation of the separator handle turns a worm gear mechanism which causes the separator bowl to spin at thousands of revolutions per minute.

When spun, the heavier milk is pulled outward against the walls of the separator and the cream, which is lighter, collects in the middle. The cream and milk then flow out of separate spouts. After separation, the cream and skimming milk are mixed together in a certain ratio until the favoured fat content has been set. The ratio is depending on the product which is to be produced (low-fat milk, fullfat milk or cream). Some floor model separators were built with a swinging platform attached to the stand. The bucket for collecting the cream was put on the platform, and a much larger bucket was set on the floor to collect the milk. Some floor model separators had two swinging platforms. Smaller versions of separators were called table-top models, for small dairies with only a few cows or goats.

Gustaf de Laval's construction made it possible to start the largest separator factory in the world, Alfa Laval AB. The milk separator represented a significant industrial breakthrough in Sweden. Within the first decade of the 1900s, there were over twenty separator manufacturers in Stockholm. Separators in modified form are also used on ships to purify oil, which may have been their original use, because in its original form de Laval proposed the separator for use in his steam turbine. De Laval's turbine used mechanically lubricated journal bearings which weren't insulated from the inside of the turbine. When the steam condensed into water it contaminated the oil. To purify the oil a centrifugal separator was used, which was later adapted to the dairy industry.

The original design had a manual bowl that required manual cleaning. Most modern separators use a self-ejecting centrifuge bowl that can automatically discharge any sedimentary solids that may be present, and that allow for clean-in-place (CIP).

A distinction is made between warm milk skimming and cold milk skimming:

- Warm milk skimming separator: At first the raw milk is heated and then skimmed warm. There is a significant difference in density between cream and skimmed milk, because of the higher temperature.

- Cold milk skimming separator: Because of the lower energy, which is used, the production cost will be reduced. Also at cold temperatures, the growth of microorganisms is significantly reduced. In the USA, Mexico, Australia and New Zealand, cold milk skimming is on the rise.

Bovine Somatotropin

Bovine somatotropin is a growth hormone found in cattle. The word bovine refers to cattle, and the word somatotropin refers to the name of the hormone. Hormones are chemicals that are secreted by glands within the body. They are natural substances that affect the way the body operates. Bovine somatotropin, abbreviated as bST, is a protein hormone produced in cattle by the pituitary gland located at the base of the animals brain.

A hormone similar to bST is produced in all species of animals. This hormone is important for growth, development, and other bodily functions of all animals. In the 1930s, it was discovered that injecting bST into lactating (milk-producing) cows significantly increased milk production.

Development of bST by Scientists

Until recently, the only source of bST was from the pituitary glands of slaughtered cattle. There were only small quantities of bST available, and it was very expensive.

Now, the new science of biotechnology makes it possible to work with DNA, the part of a cell that contains the genetic information for an animal or a plant. Scientists have determined which gene in cattle controls or codes for the production of bST. They have removed this gene from cattle and inserted it into a bacterium called Escherichia coli. This bacterium, which is found in the intestinal tract of humans and animals, acts like a tiny factory and produces large amounts of bST in controlled laboratory conditions. The bST produced by the bacteria is purified and then injected into cattle.

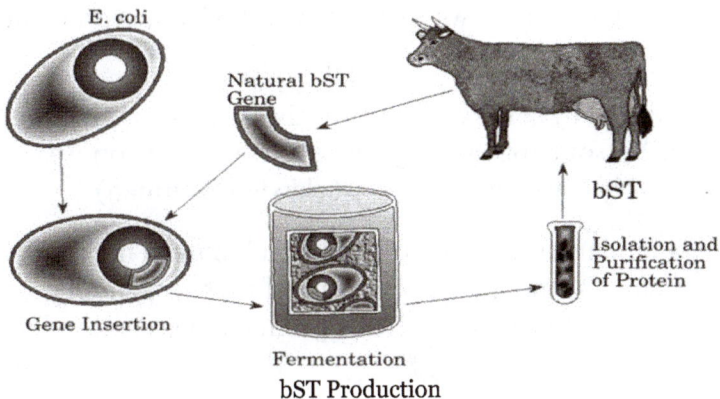

bST Production

The movement of a gene from one organism to another, in this case from the pituitary gland of a cow to a bacterial organism, is called recombinant DNA technology. Several Food and Drug Administration (FDA) approved drugs, including insulin for the treatment of diabetes and tissue plasminogen activator (TPA) for the treatment of heart attacks in people, are produced in a similar way.

Effect of bST on Milk Production

To affect a cows milk production, bST must be injected into the animal on a regular basis, similar to the way insulin must be regularly injected into people who have certain types of diabetes. Feeding bST to cows will not work. Amino acids and peptides are the building blocks of proteins.

The hormone bST is a complex protein that is immediately broken down into small, inactive amino acids and peptides and rendered ineffective when it enters a cows digestive system. How often a cow must be injected with bST will depend on whether a bST product can be developed that releases the hormone gradually over a long period of time.

Milk yields are significantly increased when cows are injected with bST, although not as much as some reports in popular newspapers and magazines suggest. The exact details of how bST increases milk production are not known, but it is thought that blood flow to the cows mammary (milk-producing) gland is increased. The blood carries an increased amount of nutrients available for milk production. More nutrients are extracted from the blood by the mammary gland, which improves efficiency of milk production. Feed efficiency (pounds of milk produced per pound of feed consumed) is improved because more milk is produced and the proportion of feed used for body maintenance is decreased. The actual amount of feed consumed by bST-treated cows increases, helping the cow meet the increased nutrient demands.

Milk production in bST-treated cows increases from 4.8 to 11.2 pounds per day. Feed efficiency improves from 2.7 to 9.3 percent (Peel, et al.). Table 1 summarizes the results of 32-week treatments of cows injected with bST in several states and foreign countries.

Misinformation provided by some groups gives the impression that there is controversy about the biology of somatotropin. However, 800 reports on 20,000 treated cows have yielded remarkably consistent results worldwide (Bauman).

Researchers have summarized several bST trials and found a milk production increase of 8.4 pounds per day (Bauman). They estimated that, depending on how the dairy operation is managed, average increased milk production is expected to range from 8.5 to 17.6 percent.

It is difficult to predict how individual cows will respond to bST. A higher response is seen when treatment is started after the cow has been producing milk for 101 days, rather than when treatment is started on days 57-100 after calving. The response o f cows treated in early lactation is less (Bauman). Cows that have had more than one calf show a greater increase in milk production than do first lactation heifers (Peel, et al.). Milk yield gradually increases for the first few days after bST treatment begins. A maximum increase is seen in about six days. To meet the needs for this increased milk production, treated cows consume from 10 to 20 percent more grain and forage.

Normally, cows reach their peak milk production 7-9 weeks after lactation begins. Milk quantity then slowly declines throughout the remainder of lactation. The ability of cows to maintain relatively high levels of milk production throughout lactation is called "persistency." The major response of cows treated with bST is a significant improvement in persistency. The normal decrease in milk yield as lactation progresses is markedly reduced. Quality of management, including health programs, milking pra ctices, nutrition, cow condition, and environmental conditions will be major factors in the response to bST.

Benefits and Risks of bST

The commercial use of bST in dairy cattle is controversial and has stirred heated debate among the dairy industy, activist groups, and consumers.

Effects on Cow Health

The physiological effects of bST treatment are the same as those seen in any high-producing cow. Nutrition, health programs, environment, and milking technique must be appropriate for the use of bST or results will be disappointing. On many farms, the management changes instituted by producers as they are preparing to use bST will probably cause a greater increase in milk production, efficiency, and profitability than actual use of bST. In the initial stages of use, producers will be encouraged to use bST on cows that have been in lactation for at least 100 days, are in good physical condition, pregnant, and are free from health problems such as mastitis or infertility.

Concern has been expressed regarding the effect of bST on reproduction. The optimum calving interval of 12-13 months may lengthen because bST can extend the time that cows efficiently produce milk. Dairy Herd Improvement Association (DHIA) records show that higher milk-producing herds have lower conception rates than lower producing herds (Ferguson and Skidmore). This negative effect on reproduction is seen in cows treated with bST and is associated with increased milk production. However, some people believe that a longer calving interval could benefit the health of bST- treated cows, since many health problems of dairy cows are associated with calving and rebreeding. The ability of a cow to reproduce is affected by her physical condition, nutrition, health, and level of milk production.

Few research studies have investigated the physiological effects of bST on the functioning of the ovaries and pituitary gland. Cows receiving dosages of bST far beyond what will be used in practice have shown an adverse effect on estrous activity (the time when an animal is capable of being bred). This effect is not seen when cows receive low to average dosages of bST. High dosages of bST are reported to increase the death rate of calf embryos, so starting a cow on bST during early pregnancy should probably be avoided (Ferguson and Skidmore). This effect is not seen at recommended dosages. The effect of bST on reproduction will have to be monitored closely in individual herds.

Several research studies have shown that bST is not associated with increased mammary infections (mastitis) (Ferguson and Skidmore). Other studies have shown an increase in mammary gland infections when bST is used, but the increase is what would be expected with increased production. The length of a cowUs gestation (pregnancy), calf birth weight, calf survival rate, and calf growth are not influenced by using the product. Some early reports indicated an increased incidence of twins, but later reports failed to confirm this.

Milk Composition and Safety

Consumer advocates and others have expressed concern about the safety of milk from bST-treated cows. All milk contains natural bST that is produced by the cow. Milk from bST-treated cows also contains the same amounts of injected bST and no differences can be measured compared to untreated cows. There are four forms of natural bST, and each has a chain of either 190 or 191 amino acids. The recombinant bST that is injected into cows has 191 amino acids. The biological activity of commercial bST is identical to naturally produced bST.

Studies indicate that both natural bST produced by the cow and bST produced by recombinant DNA techniques are immediately broken down into inactive amino acids and peptides in the digestive tract when they are consumed by humans. In contrast, steroid hormones such as estrogens, progesterones, and anabolic steroids are smaller, ring-like structures that are absorbed from the digestive tract and are biologically active in humans. This is not the case with bST in milk, whether it is produced naturally by the cow or by recombinant DNA technology (Barbano and Lynch).

Studies show that bovine somatotropin is inactive in humans. During the 1950s, natural bST produced by cows was injected into children with growth defects in an attempt to encourage growth. There was no effect, probably because the bovine somatotropin protein molecule differs from human somatotropin (human growth hormone) by about 30 percent of the amino acid sequences.

Milk composition from bST-treated cows has been thoroughly investigated (Barbano and Lynch). The characteristics of milk from bST-treated cows are within the normal range of variation of milk from untreated cows. During the first 28 days of treatment, milk fat increases and milk protein decreases slightly. After longer treatment, cows adjust their nutrient intake and the normal balance is re-established. An increase in non-protein nitrogen and whey protein and a decrease in casein have been observed aft er long-term bST administration. This difference is not always significant, and the effect on cheese yield would probably be minor, if any. One study showed a slight increase in unsaturated compared to saturated fat. The difference was small, but suggested a healthier product from bST treatment. No differences in free fatty acids have been observed. Cholesterol levels are in the range of normal milk composition. Insulin-like growth factor I increases by up to two-fold in milk from treated cows, but it is still well within the range for both bovine and human milk. No differences in flavor have been found.

The National Institute of Health has concluded that milk from bST-treated cows is essentially the same as from untreated cows, and there is no difference in safety of the products.

Economics

The potential economic effect of bST on the family dairy farm has generated heated debate. The Animal Health Institute, an organization of drug and vaccine manufacturers, maintains that the use of bST will be of equal value to any size farm (Milligan). They contend that use of the product will favor the good dairy manager, regardless of farm size.

Estimates of the effect of bST on dairy production have probably been exaggerated. The United States Department of Agriculture estimates that the use of bST could lead to a 2 to 5 percent increase in national milk production within five years, or about the increase seen yearly without the use of bST. This increase would be in addition to the normal milk production increase per cow.

In most dairy herds, bST will not be used in cows until they have been in lactation for about 100 days. It will not be used in cows with chronic health or fertility problems. It is expected that bST will be used less in heifers than in adult cows. If 50 percent of farmers adopt the use of bST, and it is used in 60 percent of the lactation days per user herd, milk production will increase about 3.5 percent (assuming an average per cow production increase of 15 percent). Many well-managed dairy herds increase per cow production more than this on an annual basis by using improved management and genetics. For most herds, a farmer who requests a thorough herd analysis by a competent nutritionist and veterinarian and then follows their recommendations will achieve a larger increase in milk production than by using bST alone.

Failure to adopt proven technology is a problem throughout the dairy industry. Almost 60 percent of cows are bred by mating to a bull, rather than by artificial insemination from proven sires with superior genetic performance. Only 50 percent of U.S. dairy producers use DHIA management information and records to improve production.

The government milk price support system tends to make prediction of the effect of bST on milk prices difficult. It is true, however, that efficient managers in areas of the country with higher milk prices benefit more from application of technology and increased production.

It is argued that large commercial dairy operations can begin using new technologies such as bST more easily, rapidly, and efficiently than smaller operations. Sophisticated record keeping and division of labor may make timely injection of cows with bST more feasible for these larger operations. Other demands on the time and management skills of typical Midwest dairy producers who have diversified farming operations may make new technology more difficult to implement.

Others argue that smaller producers with direct owner control of the herd can manage individual cows better and will see a greater production increase from the use of bST. It is not automatically true that larger herds are better managed and, therefore, will benefit more from bST.

There is no question that consumer loss of confidence in the quality of milk produced by using bST, whether the reasons are logical or not, would reduce milk consumption and have a negative economic impact on the dairy industry. This is a major concern of dairy producers. Relative to milk quality, bST appears to be neutral. It neither improves nor harms quality. Consumers would gain with bST technology since milk production costs may decline due to improved efficiency. The ultimate effect of use of bST on consumption is unknown.

Regulation of bST

The United States Food and Drug Administration (FDA) is responsible for regulating the use of bST since it is an animal drug and because milk and meat are food products. Several commercial companies have submitted data to the FDA, asking for approval to use bST in dairy cows to increase milk production.

Before approval of bST for use in dairy herds, the FDA allowed the consumption of milk and meat from animals that received bST as part of the experimental testing process. Such approval is often granted during the process of license approval of animal products. Milk from treated cows has been judged safe because bST is biologically inactive in humans and is a protein hormone that is digested and destroyed by gastric enzymes when it is consumed. Each company seeking approval for bST has to demonstrate that bST has zero biological activity in milk when it is consumed. The FDA has found no pertinent information indicating that food derived from bST-treated cows is unsafe. (Review). It is not required that producers withhold milk from the market for a certain period of time after test herd cows have been treated with bST.

Federal law prohibits the social and economic effects of a product from affecting the FDA's decision whether to approve its use or not. The FDA must determine if a product is safe, pure, potent, and effective. Producers can decide whether a product is economical or useful. Approval by the FDA does not mean that a product must be used, but only that it can be used, if desired.

Testing Required by the FDA

Before any new product can be approved, companies must demonstrate its effectiveness under actual use conditions in several geographic locations. Fifty cows per herd are required for bST approval. Three dosages of bST were used for the studies submitted to the FDA. The quality control of bST used in the test herds was monitored and all procedures to be used were approved by the FDA before the testing began. The majority of the tests in the approval process were performed by independent scientists at university laboratories and farms or in commercial herds.

Cows were injected with bST at various times during the lactation period. The effectiveness of the drug and its safety for the first and later lactation periods were monitored. Milk yield was calculated on a 3.5 percent fat basis. Milk composition, including fat, crude protein, lactose, calcium, and phosphorus, were measured about once per week. Daily feed intake was measured in the test herds. Body condition and health were monitored throughout the studies. The effect of bST treatment on reproducti on was evaluated, including breeding cycles, conception rates, number of breedings per conception, length of time from calving to the next conception, abortions, incidence of twins, calving difficulties, and stillbirths. The weight, growth, and health of calves during the first four weeks of life were monitored. Monthly somatic cell counts, as a measure of mastitis, were required. The sites where bST was injected were monitored for any signs of adverse reactions.

To evaluate safety, companies had to use one, three, and five times the expected dosage level of bST for two consecutive lactations in one of their test herds. Heifers born to treated cows were raised through breeding age and monitored for abnormalities. Companies seeking approval for bST were also required to prove that its use was not harmful to the environment.

First bST Product Approval Granted by FDA

On November 5, 1993, the FDA announced approval of a bST product, the animal drug sometribove, for increasing milk production in dairy cows. The Monsanto Company of St. Louis, Missouri, developed the drug. However, the drug could not be used immediately due to a 90-day moratorium imposed by Congress during the summer of 1993. The moratorium was designed to give the White House Office of Management and Budget time to study possible consumer reaction and the drug's impact on the dairy industry.

The FDA approval also carried with it some provisions to deal with antibiotic residue concerns. In September 1992, the General Accounting Office reported that the FDA had found evidence in submitted clinical trials that bST-treated cows have a slightly increased incidence of mastitis. This report raised concerns that antibiotic treatments for mastitis could lead to increased antibiotic residues in milk. States require milk to be tested for drug residues. Milk found to have unsafe levels of residues must be discarded.

Although an FDA advisory committee concluded in March 1993 that adequate safeguards exist to prevent unsafe levels of antibiotic residues from entering the milk supply, additional steps were taken to ensure that any unsafe residues in the milk of bST-treated cows are detected before the milk or its products are marketed.

According to a news release issued by the U.S. Department of Health and Human Services (HHS News), Monsanto agreed to a post-approval monitoring program that includes:

- A two-year tracking system of milk production and drug residues in 21 top dairy states that will periodically compare the amount of milk discarded after bST is marketed to the amount discarded prior to approval.

- A 12-month comparison of the proportion of milk discarded due to positive drug tests between bST-treated and untreated herds.

- A reporting system to monitor all bST use and follow up on all complaints.

- The use of sometribove in 24 commercial dairy herds will be specifically monitored for mastitis, animal drug use, and the resulting loss of milk.

- The FDA has concluded that it has no legal basis to require special labeling of food products derived from bST-treated cows. Food companies may voluntarily label their products, provided the information is truthful and not misleading to consumers.

Controversies Concerning the use of bST

The most intense controversy surrounding approval of bST for use in dairy cows has occurred in major dairy producing states in the Great Lakes and New England areas. Representatives of the dairy industry are concerned about the ultimate economic effect on producers.

Consumer and environmental advocacy groups have expressed opposition based on concerns about milk quality and the use of biotechnology in general.

Dairy Producers

Some producers are afraid that they will not be able to keep up with new technologies and they will suffer economically as a result. Others feel that a product such as bST will work to the disadvantage of producers in the Great Lakes States and the Northeast.

Natural resistance to new technology adoption and a fear of genetic engineering techniques cause some producers to resist the approval and use of bST.

Special Interest Groups

Activist groups with a variety of agendas and motives have addressed the bST issue. Some have stated that milk from treated cows may not be safe after all, and more testing is needed. Others see this as a scare tactic to delay or block the use of bST and undermine consumer confidence in milk from bST-treated cows. Some animal rights groups see the use of animals for food, under any circumstances, as inhumane or a violation of those animals' "rights". Others have stated that cows have a right not to be injected with bST.

Others

Other opposing arguments state that the FDA does no independent testing of its own, but only monitors the studies of the companies seeking approval. The persistent oversupply of milk and dairy products has also been cited as a reason to block the use of bST. Some dairy farmers oppose the use of bST but feel they would have no choice but to use the product in their own herd in order to stay competitive if bST came into general use (McDermott).

Milk Cooling Tanks

Dairy farms rely on highly efficient cooling of the milk to keep the milk at a consistent temperature of about 4°C in the milk tanks until the milk is collected for further processing.

Milk is stored at the farms in either closed or open milk tanks.To maintain the quality of the milk it is quickly cooled from 38°C leaving the cow to 4°C in the milk tank. The milk tank is typically equipped with a mixer to accelerate the cooling process and homogenize the milk.

In the Danfoss range of solutions for efficient cooling of milk tanks you will find solutions suited for any tank size, any evaporation temperature and for all types of refrigerants, including low GWP refrigerants. The accurate temperature control of the Danfoss components also ensures that the milk does not freeze to avoid depleting of the fresh milk quality. Our solutions are tailored to comply with national/regional legislation.

Milk Cooling in Dairy Farms

Milk must be cooled from 98 degrees F. (37 degrees C.) to storage temperature, typically about 38 degrees F., to preserve its quality. The cooling process involves removing 56

BTUs of energy from each pound of milk (27 kilojoule per kg). Typically, a refrigeration system does this by using a special refrigerant fluid to remove heat from the milk and "reject" the heat (usually) into the outside air.

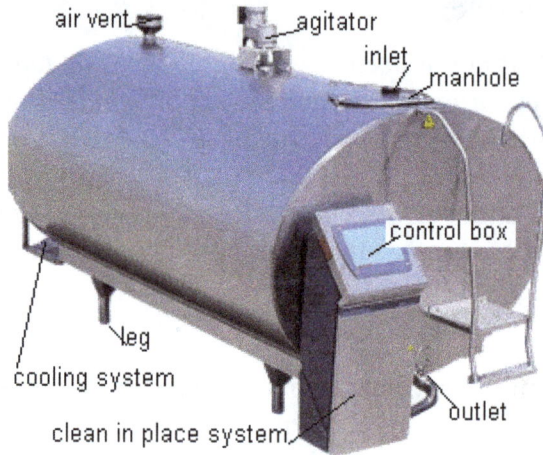

Bulk Milk Cooling Tank

The basic refrigeration system is made up of a refrigerated bulk tank, a refrigeration compressor unit and an air-cooled condenser unit. There are several technologies that can be added to the milk cooling systems on dairy farms to reduce the refrigeration requirements or to capture waste heat for pre-heating water:

- Refrigeration heat recovery (RHR) units will make a refrigeration system more efficient by collecting heat that would normally be wasted to the air and using it for water heating. An RHR unit captures heat from the system refrigerant and transfers it to water, preheating it before it enters a water heater.

- Scroll compressors are 15 to 20% more efficient than traditional reciprocation compressors yet have fewer moving parts and are only slightly more expensive than reciprocating compressors. Scroll compressors have been used in the dairy industry with good results for over 15 years. If you are purchasing a new bulk tank or replacing a failed reciprocating compressor, you should specify that the compressors be a scroll type. The additional investment is a modest cost for the improvement in efficiency.

- Well Water Precoolers are heat exchangers that use well water to cool the milk before it reaches the bulk tank. Properly sized, they can reduce milk cooling costs by up to 60%, assuming 55°F well water. Undersized water lines and water system capacity are the two largest reasons that precoolers do not perform up to their potential. Caution: If an RHR unit is being used on the dairy, an energy audit should be done, as precoolers and RHR units are competing technologies. It is usually more cost effective to maximize water heating with an RHR than precooler.

It is also important to keep a farm's refrigeration system clean and well maintained. Dirty coils or low refrigerant pressures will reduce efficiency and increase operating costs. Many systems have a "watch glass" that can be used to determine if the refrigerant needs to be recharged – if the fluid in the watch glass is bubbly instead of clear, it is time to call your refrigeration technician and schedule a service.

References

- Hutjens, Mike (24 February 1999). "Managing the Transition Cow". University of Illinois Extension. Retrieved 30 January 2011

- Devries, & Von Keyserlingk. (2005). Time of Feed Delivery Affects the Feeding and Lying Patterns of Dairy Cows. Journal of Dairy Science, 88(2), 625–631

- Moran, John; Doyle, Rebecca (2015). Cow Talk. CSIRO Publishing. ISBN 9781486301614

- U.S. Department of Agriculture, National Agriculture Statistics Service (March 2009). "Milk Cows and Production Estimates 2003–2007" (PDF). Retrieved 30 January 2011.

- Knaus (2009). "Dairy cows trapped between performance demands and adaptability". Journal of the Science of Food and Agriculture. 89 (7): 1107–1114. doi:10.1002/jsfa.3575

Chapter 5

Horses: Breeding and Management

Horse breeding is a process of human-directed selective breeding of horses in order to achieve the desired characteristics in the horse. It is done particularly in purebred domesticated horses. Horse management refers to the attention and care of horses in order to ensure their optimal health and long life. This chapter addresses some of the basic practices of husbandry for horses and horse care, such as horse grooming, horse sheath cleaning, hoof care, the correct use of stable bandage, besides others.

Horse Breeding

Horse breeding refers to reproduction in horses, and particularly the human-directed process of planned mating of animals.

While feral and wild horses breed successfully without human assistance, it can be beneficial to domesticated horses.

Humans can increase the chances of conception, a successful pregnancy, and successful foaling.

The male parent of a horse, a stallion, is commonly known as the sire and the female parent, the mare, is called the dam.

Both are genetically important, as each parent provides 50% of the genetic makeup of the ensuing offspring, called a foal. (Contrary to popular misuse, the word "colt" refers to a young male horse only.) Though many amateur horse owners may simply breed a family mare to a local stallion in order to produce a companion animal, most professional breeders use selective breeding to produce individuals of a given phenotype, or breed.

Alternatively, a breeder could, using individuals of differing phenotypes, create a new breed with specific characteristics.

Terminology

The male parent of a horse, a stallion, is commonly known as the sire and the female parent, the mare, is called the dam. Both are genetically important, as each parent provides half of the genetic makeup of the ensuing offspring, called a foal. Contrary to popular misuse, "colt" refers to a young male horse only; "filly" is a young female. Though many horse owners may simply breed a family mare to a local stallion in order to produce a companion animal, most professional breeders use selective breeding to produce individuals of a given phenotype, or breed. Alternatively, a breeder could, using individuals of differing phenotypes, create a new breed with specific characteristics.

A horse is "bred" where it is foaled (born). Thus a foal conceived in England but foaled in the United States is regarded as being bred in the US. In some cases, most notably in the Thoroughbred breeding industry, American- and Canadian-bred horses may also be described by the state or province in which they are foaled. Some breeds denote the country, or state, where conception took place as the origin of the foal.

Similarly, the "breeder", is the person who owned or leased the mare at the time of foaling. That individual may not have had anything to do with the mating of the mare. It is important to review each breed registry's rules to determine which applies to any specific foal.

In the horse breeding industry, the term "half-brother" or "half-sister" only describes horses which have the same dam, but different sires. Horses with the same sire but different dams are simply said to be "by the same sire", and no sibling relationship is implied. "Full" (or "own") siblings have both the same dam and the same sire. The terms paternal half-sibling, and maternal half-sibling are also often used. Three-quarter siblings are horses out of the same dam, and are by sires that are either half-brothers (i.e. same dam) or who are by the same sire.

Thoroughbreds and Arabians are also classified through the "distaff" or direct female line, known as their "family" or "tail female" line, tracing back to their taproot foundation bloodstock or the beginning of their respective stud books. The female line of descent always appears at the bottom of a tabulated pedigree and is therefore often known as the bottom line. In addition, the maternal grandfather of a horse has a special term: damsire.

"Linebreeding" technically is the duplication of fourth generation or more distant ancestors. However, the term is often used more loosely, describing horses with duplication of ancestors closer than the fourth generation. It also is sometimes used as a euphemism for the practice of inbreeding, a practice that is generally frowned upon by horse breeders, though used by some in an attempt to fix certain traits.

Estrous Cycle of the Mare

The estrous cycle (also spelled oestrous) controls when a mare is sexually receptive toward a stallion, and helps to physically prepare the mare for conception. It generally occurs during the spring and summer months, although some mares may be sexually receptive into the late fall, and is controlled by the photoperiod (length of the day), the cycle first triggered when the days begin to lengthen. The estrous cycle lasts about 19–22 days, with the average being 21 days. As the days shorten, the mare returns to a period when she is not sexually receptive, known as anestrus. Anestrus – occurring in the majority of, but not all, mares – prevents the mare from conceiving in the winter months, as that would result in her foaling during the harshest part of the year, a time when it would be most difficult for the foal to survive.

Stallion checking a mare in estrus. The mare welcomes the stallion
by lowering her rear and lifting her tail

This cycle contains 2 phases:

- Estrus, or Follicular, phase: 5–7 days in length, when the mare is sexually receptive to a stallion. Estrogen is secreted by the follicle. Ovulation occurs in the final 24–48 hours of estrus.

- Diestrus, or Luteal, phase: 14–15 days in length, the mare is not sexually receptive to the stallion. The corpus luteum secretes progesterone.

Depending on breed, on average, 16% of mares have double ovulations, allowing them to twin, though this does not affect the length of time of estrus or diestrus.

Effects on the Reproductive System during the Estrous Cycle

Changes in hormone levels can have great effects on the physical characteristics of the reproductive organs of the mare, thereby preparing, or preventing, her from conceiving.

- Uterus: increased levels of estrogen during estrus cause edema within the uterus, making it feel heavier, and the uterus loses its tone. This edema decreases following ovulation, and the muscular tone increases. High levels of progesterone do not cause edema within the uterus. The uterus becomes flaccid during anestrus.

- Cervix: the cervix starts to relax right before estrus occurs, with maximal relaxation around the time of ovulation. The secretions of the cervix increase. High progesterone levels (during diestrus) cause the cervix to close and become toned.

- Vagina: the portion of the vagina near the cervix becomes engorged with blood right before estrus. The vagina becomes relaxed and secretions increase.

- Vulva: relaxes right before estrus begins. Becomes dry, and closes more tightly, during diestrus.

Hormones Involved in the Estrous Cycle, during Foaling, and after Birth

The cycle is controlled by several hormones which regulate the estrous cycle, the mare's behavior, and the reproductive system of the mare. The cycle begins when the increased day length causes the pineal gland to reduce the levels of melatonin, thereby allowing the hypothalamus to secrete GnRH.

- GnRH (Gonadotropin releasing hormone): secreted by the hypothalamus, causes the pituitary to release two gonadotrophins: LH and FSH.

- LH (Luteinizing hormone): levels are highest 2 days following ovulation, then slowly decrease over 4–5 days, dipping to their lowest levels 5–16 days after ovulation. Stimulates maturation of the follicle, which then in turn secretes estrogen. Unlike most mammals, the mare does not have an increase of LH right before ovulation.

- FSH (Follicle-stimulating hormone): secreted by the pituitary, causes the ovarian follicle to develop. Levels of FSH rise slightly at the end of estrus, but have their highest peak about 10 days before the next ovulation. FSH is inhibited by inhibin, at the same time LH and estrogen levels rise, which prevents immature follicles from continuing their growth. Mares may however have multiple FSH waves during a single estrous cycle, and diestrus follicles resulting from a diestrus FSH wave are not uncommon, particularly in the height of the natural breeding season.

- Estrogen: secreted by the developing follicle, it causes the pituitary gland to secrete more LH (therefore, these 2 hormones are in a positive feedback loop). Additionally, it causes behavioral changes in the mare, making her more receptive toward the stallion, and causes physical changes in the cervix, uterus, and vagina to prepare the mare for conception. Estrogen peaks 1–2 days before ovulation, and decreases within 2 days following ovulation.

- Inhibin: secreted by the developed follicle right before ovulation, "turns off" FSH, which is no longer needed now that the follicle is larger.

- Progesterone: prevents conception and decreases sexual receptibility of the mare to the stallion. Progesterone is therefore lowest during the estrus phase, and increases during diestrus. It decreases 12–15 days after ovulation, when the corpus luteum begins to decrease in size.

- Prostaglandin: secreted by the endrometrium 13–15 days following ovulation, causes luteolysis and prevents the corpus luteum from secreting progesterone

- eCG – equine chorionic gonadotropin – also called PMSG (pregnant mare serum gonadotropin): chorionic gonadotropins secreted if the mare conceives. First secreted by the endometrial cups around the 36th day of gestation, peaking around day 60, and decreasing after about 120 days of gestation. Also help to stimulate the growth of the fetal gonads.

- Prolactin: stimulates lactation.

- Oxytocin: stimulates the uterus to contract.

Breeding and Gestation

While horses in the wild mate and foal in mid to late spring, in the case of horses domestically bred for competitive purposes, especially horse racing, it is desirable that they be born as close to January 1 in the northern hemisphere or August 1 in the southern hemisphere as possible, so as to be at an advantage in size and maturity when competing against other horses in the same age group. When an early foal is desired, barn managers will put the mare "under lights" by keeping the barn lights on in the winter to simulate a longer day, thus bringing the mare into estrus sooner than she would in nature. Mares signal estrus and ovulation by urination in the presence of a stallion, raising the tail and revealing the vulva. A stallion, approaching with a high head, will usually nicker, nip and nudge the mare, as well as sniff her urine to determine her readiness for mating.

Once fertilized, the oocyte (egg) remains in the oviduct for approximately 5.5 more days, and then descends into the uterus. The initial single cell combination is already dividing and by the time of entry into the uterus, the egg might have already reached the blastocyst stage.

The gestation period lasts for about eleven months, or about 340 days (normal average range 320–370 days). During the early days of pregnancy, the conceptus is mobile, moving about in the uterus until about day 16 when "fixation" occurs. Shortly after fixation, the embryo proper (so called up to about 35 days) will become visible on trans-rectal ultrasound (about day 21) and a heartbeat should be visible by about day 23. After the formation of the endometrial cups and early placentation is initiated (35–40 days of gestation) the terminology changes, and the embryo is referred to as a fetus. True implantation – invasion into the endometrium of any sort – does not occur until about day 35 of pregnancy with the formation of the endometrial cups, and true placentation (formation of the placenta) is not initiated until about day 40-45 and not completed until about 140 days of pregnancy. The fetus sex can be determined by day 70 of the gestation using ultrasound. Halfway through gestation the fetus is the size of between a rabbit and a beagle. The most dramatic fetal development occurs in the last 3 months of pregnancy when 60% of fetal growth occurs.

Colts are carried on average about 4 days longer than fillies.

Care of the Pregnant Mare

Domestic mares receive specific care and nutrition to ensure that they and their foals are healthy. Mares are given vaccinations against diseases such as the Rhinopneumonitis (EHV-1) virus (which can cause abortions) as well as vaccines for other conditions that may occur in a given region of the world. Pre-foaling vaccines are recommended 4–6 weeks prior to foaling to maximize the immunoglobulin content of the colostrum in the first milk. Mares are dewormed a few weeks prior to foaling, as the mare is the primary source of parasites for the foal.

Mares can be used for riding or driving during most of their pregnancy. Exercise is healthy, though should be moderated when a mare is heavily in foal. Exercise in excessively high temperatures has been suggested as being detrimental to pregnancy maintenance during the embryonic period; however ambient temperatures encountered during the research were in the region of 100 degrees F and the same results may not be encountered in regions with lower ambient temperatures.

During the first several months of pregnancy, the nutritional requirements do not increase significantly since the rate of growth of the fetus is very slow. However, during this time, the mare may be provided supplemental vitamins and minerals, particularly if forage quality is questionable. During the last 3–4 months of gestation, rapid growth of the fetus increases the mare's nutritional requirements. Energy requirements during these last few months, and during the first few months of lactation are similar to those of a horse in full training. Trace minerals such as copper are extremely important, particularly during the tenth month of pregnancy, for proper skeletal formation. Many feeds designed for pregnant and lactating mares provide the careful balance required of increased protein, increased calories through extra fat as well as vitamins and minerals. Overfeeding the pregnant mare, particularly during early gestation, should be avoided, as excess weight may contribute to difficulties foaling or fetal/foal related problems.

Foaling

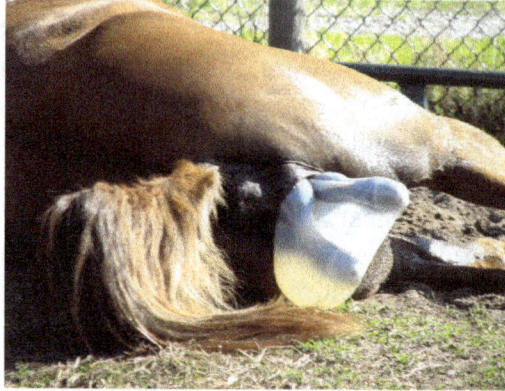

A mare in the early stages of labor

Mares due to foal are usually separated from other horses, both for the benefit of the mare and the safety of the soon-to-be-delivered foal. In addition, separation allows the mare to be monitored more closely by humans for any problems that may occur while giving birth. In the northern hemisphere a special foaling stall that is large and clutter free is frequently used, particularly by major breeding farms. Originally, this was due in part to a need for protection from the harsh winter climate present when mares foal early in the year, but even in moderate climates, such as Florida, foaling stalls are still common because they allow closer monitoring of mares. Smaller breeders often use a small pen with a large shed for foaling, or they may remove a wall between two box stalls in a small barn to make a large stall. In the milder climates seen in much of the southern hemisphere, most mares foal outside, often in a paddock built specifically for foaling, especially on the larger stud farms. Many stud farms worldwide employ technology to alert human managers when the mare is about to foal, including webcams, closed-circuit television, or assorted types of devices that alert a handler via a remote alarm when a mare lies down in a position to foal.

On the other hand, some breeders, particularly those in remote areas or with extremely large numbers of horses, may allow mares to foal out in a field amongst a herd, but may also see higher rates of foal and mare mortality in doing so.

Most mares foal at night or early in the morning, and prefer to give birth alone when possible. Labor is rapid, often no more than 30 minutes, and from the time the feet of the foal appear to full delivery is often only about 15 to 20 minutes. Once the foal is born, the mare will lick the newborn foal to clean it and help blood circulation. In a very short time, the foal will attempt to stand and get milk from its mother. A foal should stand and nurse within the first hour of life.

To create a bond with her foal, the mare licks and nuzzles the foal, enabling her to distinguish the foal from others. Some mares are aggressive when protecting their foals, and may attack other horses or unfamiliar humans that come near their newborns.

After birth, a foal's navel is dipped in antiseptic to prevent infection, it is sometimes given an enema to help clear the meconium from its digestive tract, and the newborn is monitored to ensure that it stands and nurses without difficulty. While most horse births happen without complications, many owners have first aid supplies prepared and a veterinarian on call in case of a birthing emergency. People who supervise foaling should also watch the mare to be sure that she passes the placenta in a timely fashion, and that it is complete with no fragments remaining in the uterus, where retained fetal membranes could cause a serious inflammatory condition (endometritis) and/or infection. If the placenta is not removed from the stall after it is passed, a mare will often eat it, an instinct from the wild, where blood would attract predators.

Foal Care

A foal with its mother, or dam

Foals develop rapidly, and within a few hours a wild foal can travel with the herd. In domestic breeding, the foal and dam are usually separated from the herd for a while, but within a few weeks are typically pastured with the other horses. A foal will begin to eat hay, grass and grain alongside the mare at about 4 weeks old; by 10–12 weeks the foal requires more nutrition than the mare's milk can supply. Foals are typically weaned at 4–8 months of age, although in the wild a foal may nurse for a year.

How Breeds Develop

Beyond the appearance and conformation of a specific type of horse, breeders aspire to improve physical performance abilities. This concept, known as matching "form to function," has led to the development of not only different breeds, but also families or bloodlines within breeds that are specialists for excelling at specific tasks.

For example, the Arabian horse of the desert naturally developed speed and endurance to travel long distances and survive in a harsh environment, and domestication by humans added a trainable disposition to the animal's natural abilities. In the meantime, in northern Europe, the locally adapted heavy horse with a thick, warm coat was domesticated and put to work as a farm animal that could pull a plow or wagon. This animal was later adapted through selective breeding to create a strong but rideable animal suitable for the heavily armored knight in warfare.

Then, centuries later, when people in Europe wanted faster horses than could be produced from local horses through simple selective breeding, they imported Arabians and other oriental horses to breed as an outcross to the heavier, local animals. This led to the development of breeds such as the Thoroughbred, a horse taller than the Arabian and faster over the distances of a few miles required of a European race horse or light cavalry horse. Another cross between oriental and European horses produced the Andalusian, a horse developed in Spain that was powerfully built, but extremely nimble and capable of the quick bursts of speed over short distances necessary for certain types of combat as well as for tasks such as bullfighting.

Later, the people who settled the Americas needed a hardy horse that was capable of working with cattle. Thus, Arabians and Thoroughbreds were crossed on Spanish horses, both domesticated animals descended from those brought over by the Conquistadors, and feral horses such as the Mustangs, descended from the Spanish horse, but adapted by natural selection to the ecology and climate of the west. These crosses ultimately produced new breeds such as the American Quarter Horse and the Criollo of Argentina.

In modern times, these breeds themselves have since been selectively bred to further specialize at certain tasks. One example of this is the American Quarter Horse. Once a general-purpose working ranch horse, different bloodlines now specialize in different events. For example, larger, heavier animals with a very steady attitude are bred to give competitors an advantage in events such as team roping, where a horse has to start and stop quickly, but also must calmly hold a full-grown steer at the end of a rope. On the other hand, for an event known as cutting, where the horse must separate a cow from a herd and prevent it from rejoining the group, the best horses are smaller, quick, alert, athletic and highly trainable. They must learn quickly, have conformation that allows quick stops and fast, low turns, and the best competitors have a certain amount of independent mental ability to anticipate and counter the movement of a cow, popularly known as "cow sense."

Another example is the Thoroughbred. While most representatives of this breed are bred for horse racing, there are also specialized bloodlines suitable as show hunters or show jumpers. The hunter must have a tall, smooth build that allows it to trot and canter smoothly and efficiently. Instead of speed, value is placed on appearance and upon giving the equestrian a comfortable ride, with natural jumping ability that shows bascule and good form.

A show jumper, however, is bred less for overall form and more for power over tall fences, along with speed, scope, and agility. This favors a horse with a good galloping stride, powerful hindquarters that can change speed or direction easily, plus a good shoulder angle and length of neck. A jumper has a more powerful build than either the hunter or the racehorse.

Deciding to Breed a Horse

Breeding a horse is an endeavor where the owner, particularly of the mare, will usually need to invest considerable time and money. For this reason, a horse owner needs to consider several factors, including:

- Does the proposed breeding animal have valuable genetic qualities to pass on?
- Is the proposed breeding animal in good physical health, fertile, and able to withstand the rigors of reproduction?
- For what purpose will the foal be used?
- Is there a market for the foal in the event that the owner does not wish to keep the foal for its entire life?
- What is the anticipated economic benefit, if any, to the owner of the ensuing foal?
- What is the anticipated economic benefit, if any, to the owners of the sire and dam or the foal?
- Does the owner of the mare have the expertise to properly manage the mare through gestation and parturition?
- Does the owner of the potential foal have the expertise to properly manage and train a young animal once it is born?

There are value judgements involved in considering whether an animal is suitable breeding stock, hotly debated by breeders. Additional personal beliefs may come into play when considering a suitable level of care for the mare and ensuing foal, the potential market or use for the foal, and other tangible and intangible benefits to the owner.

If the breeding endeavor is intended to make a profit, there are additional market factors to consider, which may vary considerably from year to year, from breed to breed, and by region of the world. In many cases, the low end of the market is saturated with horses, and the law of supply and demand thus allows little or no profit to be made from breeding unregistered animals or animals of poor quality, even if registered.

The minimum cost of breeding for a mare owner includes the stud fee, and the cost of proper nutrition, management and veterinary care of the mare throughout gestation, parturition, and care of both mare and foal up to the time of weaning. Veterinary expenses may be higher if specialized reproductive technologies are used or health complications occur.

Making a profit in horse breeding is often difficult. While some owners of only a few horses may keep a foal for purely personal enjoyment, many individuals breed horses in hopes of making some money in the process.

A rule of thumb is that a foal intended for sale should be worth three times the cost of the stud fee if it were sold at the moment of birth. From birth forward, the costs of care and training are added to the value of the foal, with a sale price going up accordingly. If the foal wins awards in some form of competition, that may also enhance the price.

On the other hand, without careful thought, foals bred without a potential market for them may wind up being sold at a loss, and in a worst-case scenario, sold for "salvage" value—a euphemism for sale to slaughter as horsemeat.

Therefore, a mare owner must consider their reasons for breeding, asking hard questions of themselves as to whether their motivations are based on either emotion or profit and how realistic those motivations may be.

Choosing Breeding Stock

A stallion with a proven competition record is one criterion for being a suitable sire

The stallion should be chosen to complement the mare, with the goal of producing a foal that has the best qualities of both animals, yet avoids having the weaker qualities of either parent. Generally, the stallion should have proven himself in the discipline or sport the mare owner wishes for the "career" of the ensuing foal. Mares should also have a competition record showing that they also have suitable traits, though this does not happen as often.

Some breeders consider the quality of the sire to be more important than the quality of the dam. However, other breeders maintain that the mare is the most important parent. Because stallions can produce far more offspring than mares, a single stallion can have a greater overall impact on a breed. However, the mare may have a greater influence on an individual foal because its physical characteristics influence the developing foal in the womb and the foal also learns habits from its dam when young. Foals may also learn the "language of intimidation and submission" from their dam, and this imprinting may affect the foal's status and rank within the herd. Many times, a mature horse will achieve status in a herd similar to that of its dam; the offspring of dominant mares become dominant themselves.

A purebred horse is usually worth more than a horse of mixed breeding, though this matters more in some disciplines than others. The breed of the horse is sometimes secondary when breeding for a sport horse, but some disciplines may prefer a certain breed or a specific phenotype of horse. Sometimes, purebred bloodlines are an absolute requirement: For example, most racehorses in the world must be recorded with a breed registry in order to race.

Bloodlines are often considered, as some bloodlines are known to cross well with others. If the parents have not yet proven themselves by competition or by producing quality offspring, the bloodlines of the horse are often a good indicator of quality and possible strengths and weaknesses. Some bloodlines are known not only for their athletic ability, but could also carry a conformational or genetic defect, poor temperament, or for a medical problem. Some bloodlines are also fashionable or otherwise marketable, which is an important consideration should the mare owner wish to sell the foal.

Horse breeders also consider conformation, size and temperament. All of these traits are heritable, and will determine if the foal will be a success in its chosen discipline. The offspring, or "get", of a stallion are often excellent indicators of his ability to pass on his characteristics, and the particular traits he actually passes on. Some stallions are fantastic performers but never produce offspring of comparable quality. Others sire fillies of great abilities but not colts. At times, a horse of mediocre ability sires foals of outstanding quality.

Mare owners also look into the question of if the stallion is fertile and has successfully "settled" (i.e. impregnated) mares. A stallion may not be able to breed naturally, or old age may decrease his performance. Mare care boarding fees and semen collection fees can be a major cost.

Costs Related to Breeding

Breeding a horse can be an expensive endeavor, whether breeding a backyard competition horse or the next Olympic medalist. Costs may include:

- The stud and booking fee
- Fees for collecting, handling, and transporting semen (if AI is used and semen is shipped)

- Mare exams: to determine if she is healthy enough to breed, to determine when she ovulates, and (if AI is used) to inseminate her

- Mare transport, care, and board if the mare is bred live cover at the stallion's residence

- Veterinary bills to keep the pregnant mare healthy while in foal

- Possible veterinary bills during pregnancy or foaling should something go wrong

- Veterinary bills for the foal for its first exam a few days following foaling

Stud fees are determined by the quality of the stallion, his performance record, the performance record of his get (offspring), as well as the sport and general market that the animal is standing for.

The highest stud fees are generally for racing Thoroughbreds, which may charge from two to three thousand dollars for a breeding to a new or unproven stallion, to several hundred thousand dollars for a breeding to a proven producer of stakes winners. Stallions in other disciplines often have stud fees that begin in the range of $1,000 to $3,000, with top contenders who produce champions in certain disciplines able to command as much as $20,000 for one breeding. The lowest stud fees to breed to a grade horse or an animal of low-quality pedigree may only be $100–$200, but there are trade-offs: the horse will probably be unproven, and likely to produce lower-quality offspring than a horse with a stud fee that is in the typical range for quality breeding stock.

As a stallion's career, either performance or breeding, improves, his stud fee tends to increase in proportion. If one or two offspring are especially successful, winning several stakes races or an Olympic medal, the stud fee will generally greatly increase. Younger, unproven stallions will generally have a lower stud fee earlier on in their careers.

To help decrease the risk of financial loss should the mare die or abort the foal while pregnant, many studs have a live foal guarantee (LFG) – also known as "no foal, free return" or "NFFR" - allowing the owner to have a free breeding to their stallion the next year. However, this is not offered for every breeding.

Covering the Mare

There are two general ways to "cover" or breed the mare:

- Live cover: the mare is brought to the stallion's residence and is covered "live" in the breeding shed. She may also be turned out in a pasture with the stallion for several days to breed naturally ('pasture bred'). The former situation is often preferred, as it provides a more controlled environment, allowing the breeder to ensure that the mare was covered, and places the handlers in a position to remove the horses from one another should one attempt to kick or bite the other.

- Artificial Insemination (AI): the mare is inseminated by a veterinarian or an equine reproduction manager, using either fresh, cooled or frozen semen.

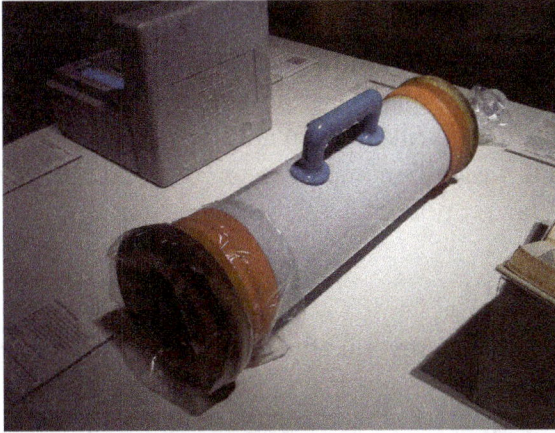

An artificial vagina, used to collect semen

After the mare is bred or artificially inseminated, she is checked using ultrasound 14–16 days later to see if she "took", and is pregnant. A second check is usually performed at 28 days. If the mare is not pregnant, she may be bred again during her next cycle.

It is considered safe to breed a mare to a stallion of much larger size. Because of the mare's type of placenta and its attachment and blood supply, the foal will be limited in its growth within the uterus to the size of the mare's uterus, but will grow to its genetic potential after it is born. Test breedings have been done with draft horse stallions bred to small mares with no increase in the number of difficult births.

Live Cover

When breeding live cover, the mare is usually boarded at the stud. She may be "teased" several times with a stallion that will not breed to her, usually with the stallion being presented to the mare over a barrier. Her reaction to the teaser, whether hostile or passive, is noted. A mare that is in heat will generally tolerate a teaser (although this is not always the case), and may present herself to him, holding her tail to the side. A veterinarian may also determine if the mare is ready to be bred, by ultrasound or palpating daily to determine if ovulation has occurred. Live cover can also be done in liberty on a paddock or on pasture, although due to safety and efficacy concerns, it is not common at professional breeding farms.

When it has been determined that the mare is ready, both the mare and intended stud will be cleaned. The mare will then be presented to the stallion, usually with one handler controlling the mare and one or more handlers in charge of the stallion. Multiple handlers are preferred, as the mare and stallion can be easily separated should there be any trouble.

The Jockey Club, the organization that oversees the Thoroughbred industry in the United States, requires all registered foals to be bred through live cover. Artificial insemination, listed below, is not permitted. Similar rules apply in other countries.

By contrast, the U.S. standardbred industry allows registered foals to be bred by live cover, or by artificial insemination (AI) with fresh or frozen (not dried) semen. No other artificial fertility treatment is allowed. In addition, foals bred via AI of frozen semen may only be registered if the stallion's sperm was collected during his lifetime, and used no later than the calendar year of his death or castration.

Artificial Insemination

Artificial insemination (AI) has several advantages over live cover, and has a very similar conception rate:

- The mare and stallion never have to come in contact with each other, which therefore reduces breeding accidents, such as the mare kicking the stallion.

- AI opens up the world to international breeding, as semen may be shipped across continents to mares that would otherwise be unable to breed to a particular stallion.

- A mare also does not have to travel to the stallion, so the process is less stressful on her, and if she already has a foal, the foal does not have to travel.

- AI allows more mares to be bred from one stallion, as the ejaculate may be split between mares.

- AI reduces the chance of spreading sexually transmitted diseases between mare and stallion.

- AI allows mares or stallions with health issues, such as sore hocks which may prevent a stallion from mounting, to continue to breed.

- Frozen semen may be stored and used to breed mares even after the stallion is dead, allowing his lines to continue. However, the semen of some stallions does not freeze well. Some breed registries may not permit the registration of foals resulting from the use of frozen semen after the stallion's death, although other large registries accept such usage and provide registrations. The overall trend is toward permitting use of frozen semen after the death of the stallion.

A stallion is usually trained to mount a phantom (or dummy) mare, although a live mare may be used, and he is most commonly collected using an artificial vagina (AV) which is heated to simulate the vagina of the mare. The AV has a filter and collection area at one end to capture the semen, which can then be processed in a lab. The semen may be chilled or frozen and shipped to the mare owner or used to breed mares "on-farm". When the mare is in heat, the person inseminating introduces the semen directly into her uterus using a syringe and pipette.

Advanced Reproductive Techniques

The Thoroughbred industry does not allow AI or embryo transplant

Often an owner does not want to take a valuable competition mare out of training to carry a foal. This presents a problem, as the mare will usually be quite old by the time she is retired from her competitive career, at which time it is more difficult to impregnate her. Other times, a mare may have physical problems that prevent or discourage breeding. However, there are now several options for breeding these mares. These options also allow a mare to produce multiple foals each breeding season, instead of the usual one. Therefore, mares may have an even greater value for breeding.

- Embryo Transfer: This relatively new method involves flushing out the mare's fertilized embryo a few days following insemination, and transferring to a surrogate mare, which has been synchronized to be in the same phase of the estrous cycle as the donor mare.

- Gamete Intrafallopian Transfer (GIFT): The mare's ovum and the stallion's sperm are deposited in the oviduct of a surrogate dam. This technique is very useful for subfertile stallions, as fewer sperm are needed, so a stallion with a low sperm count can still successfully breed.

- Egg Transfer: An oocyte is removed from the mare's follicle and transferred into the oviduct of the recipient mare, who is then bred. This is best for mares with physical problems, such as an obstructed oviduct, that prevent breeding.

- Intracytoplasmic Sperm Injection (ICSI): Used in horses due to lack of successful co-incubation of female and male gametes in simple IVF. A plug of the zona pellucida is removed and a single sperm cell is injected into the ooplasm of the mature oocyte. An advantage of ICSI over IVF is that lower quality sperm can be used since the sperm does not have to penetrate the zona pellucida. The success rate of ICSI is 23-44% blastocyst development.

Horse Grooming

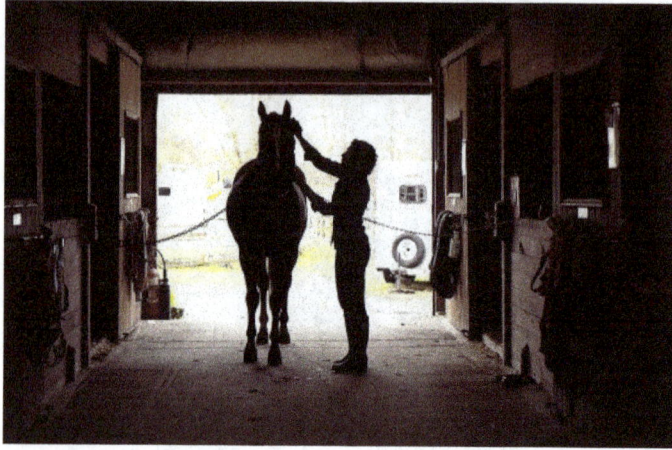

Grooming is an activity that is enjoyable for both you and your horse. It is also a good opportunity to check for injuries and irritations. Try to make grooming a daily habit. It is an absolute must before riding. Grit beneath the saddle or girth or cinch will be uncomfortable for your horse and could cause saddle or girth sores. Start from the left or right of your horse. These instructions assume you will start on the left side, but as long as you cover the whole horse, it does not matter.

Have your grooming tools arranged in a convenient, safe place. A wide bucket may be cheapest and easiest to put your brushes in, although there are lots of grooming boxes on the market that keep your tools organized and handy.

You will need:

- A curry comb or grooming mitt.

- A body brush with fairly stiff bristles.

- A mane and tail comb. Plastic causes less breakage than metal ones.

- A fine soft bristled finishing brush.

- A hoof pick.

- A clean sponge or soft cloth.

Nice to have:

- Grooming spray.

- Hoof ointment if recommended by your farrier.

- Scissors or clippers.

Don't sit your bucket or box too close to your horse where he could knock it over, or where you might trip over it as you move around your horse. Also have your horse securely and safely tied either with cross ties or with a quick release knot. If your horse seems uneasy about grooming, there may be things you can do to make it more comfortable.

Clean your Horse's or Pony's Hooves

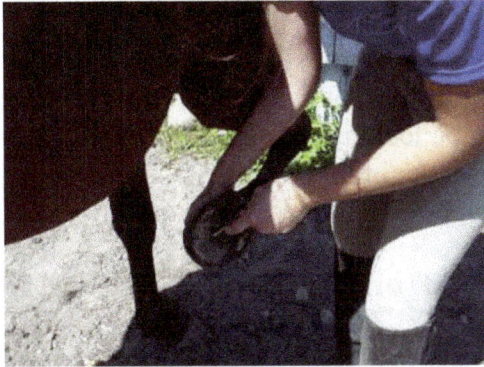

Clean out all four hooves and check for signs of injury or disease. Draw the hoof
pick back to front to clean out around the frog

Cleaning out your horse's hooves is very important. Slide your hand down the left foreleg. Squeeze the back of the leg along the tendons just above the pastern and say 'up' or 'hoof'—whatever your horse is trained to respond to. Hold the hoof and with the hoof pick pry out any dirt, manure or anything else lodged in the frog or sole of the foot. Check for any injury and signs of thrush, grease heel, or other problems. Take note of any cracks in the wall of the hoof so you can consult with your farrier as to what should be done. Gently place the foot back down on the ground and continue until all four feet are done.

Currying your Horse or Pony

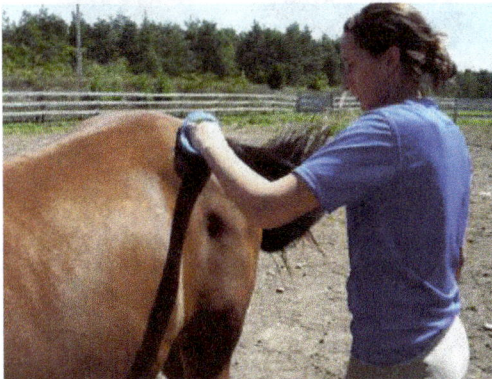

Use your curry comb or grooming mitt to dislodge the dirt in your horse's or pony's hair coat.
Use vigorous circular sweeps, being gentle over bony areas such as shoulders,
hips and legs

Starting on the left side or 'offside' use your curry comb or grooming mitt to loosen the dirt in your horse's coat. This step is where you remove any mud, grit, dust and other debris before trying to put a real shine on your horse's coat. Curry in circular sweeps all over the horse's body. Be careful over bony areas of the shoulders, hips, and legs. Use a light touch in these areas. Many horses are sensitive about having their bellies and between the back legs brushed. Be careful in these areas to use a light touch.

Some horses are more sensitive skinned than others to adjust the pressure on the brush according to what they seem to enjoy. If your horse reacts by laying back his ears or swishing his tail in agitation, he is telling you that the brushing is too vigorous. As well as currying, you will also be looking for any skin lesions or wounds. If you find anything, you'll want to assess the injury and decide if you want to treat it yourself with something out of your first aid kit, or if you need a vet to treat it.

Comb Out the Tangles from the Mane and Tail

A flowing, shiny mane and tail are a joy to behold. Get that full, healthy look by being gentle and patient as you groom your horse's mane or tail. Either with a mane comb or brush, start at the bottom of the strands and brush downwards in sections until you can smoothly comb from the top of the mane or tail, right to the bottom. When brushing the tail, stand to one side and pull the tail gently over to you. This way you are out of the way should the horse kick. A grooming spray that detangles hair is nice to have and makes brushing out the long strands easier while cleaning, shining and protecting the hair. A grooming spray may also help prevent the hairs from tangling too much between groomings. Some people like to keep their horse's tails wrapped to keep it clean and tangle-free, but I prefer to let my horse use its tail the way nature intended, to swat away flies. If your horse's mane is unruly and lies on both sides, here's how to tame it to lay flatter.

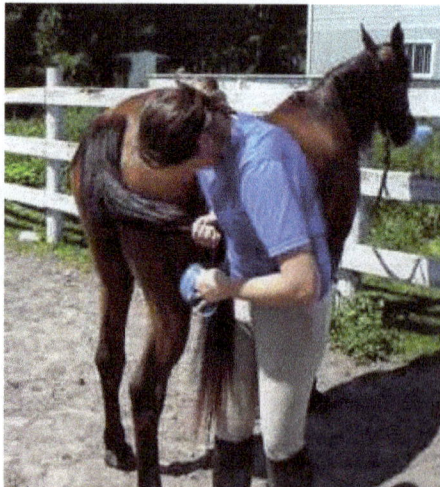

Stand to the side while gently brushing or combing through your horse's tail. Go section by section, working your way up from the bottom, brushing downwards a few inches at a time

Use the Body Brush to Whisk Away Dirt

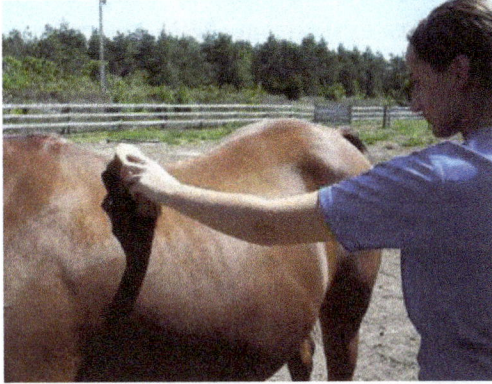

Whisk away the dirt left during currying with a stiff bristled
dandy or body brush. Katherine Blocksdorf

After currying the body to get rid of the coarser dirt, it's time to go to work with a body brush. This longer bristled, but still, a stiff brush will get rid of what the curry comb missed. With the body brush, whisk out the dirt brought to the surface. Start on one side and move around the horse brushing in sweeping strokes following the direction of the hair growth. Some people find the body brush is more useful for cleaning the legs than the curry comb. All the curves and angles on the legs make it hard to get the curry comb to do a complete job. So you can use the body brush, sometimes quite vigorously, to remove any remaining dirt. This is a good time to check for lesions and skin irritations on the legs, knees, and pasterns like small cuts and nick, or perhaps even problems like grease heel.

Using the Finishing Brush

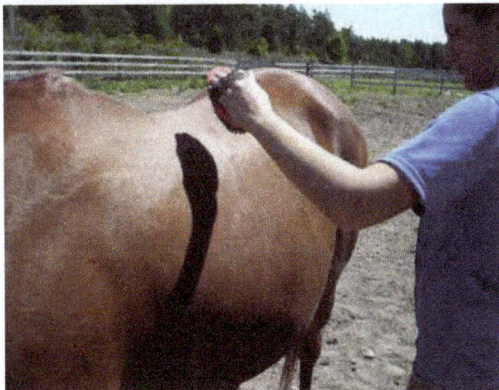

The Finishing brush makes your horse's coat sleek and glossy and removes the last traces of dust and dirt. Use long sweeping strokes over the whole body and broad areas of the face

My mother always joked that she never used a finishing brush because she never felt she was finished grooming her horse. Well, you might never want to finish, but your finishing brush will help bring out the shine on your horse's coat. A finishing brush will have shorter softer bristles and may be used on your horse's or pony's face if you don't have a special brush just for that. Gently whisk away dust from the broader areas on your horse's face, ears and throat. With sweeping strokes whisk away any dust missed by the body brush. The finer bristles help smooth out the body hair and leave your horse looking more finished and glossy. When you think you are done, you can apply a grooming spray. Grooming sprays, depending on the type, can provide sun protection, and add shine to your horse's coat, but they aren't necessary. If you plan to ride, however, then you must be aware that some products may make the hair slippery and could cause your saddle to shift. Try to avoid application to the saddle area.

Clean the Ears, Eyes, Muzzle and Dock Area

So far, you have cleaned up your horse's body, mane, and tail. Now it's time for detailing. With a damp sponge or soft cloth wipe around the horse's eyes and muzzle, and clean away any dirt or chaff. You may prefer a soft cloth as it can be more easily laundered between uses. Check your horse's eyes. A bit of tearing at the corner of the eye is not uncommon, but take note of excess tearing, redness, or swelling. Eye infections need to be treated promptly. Check ears for lodged seed heads or dirt. Some horses are fussy about having their ears handled, so go slowly and be careful not to pinch or pull hairs. Eventually, your horse may come to love having its ears groomed. When you are done with the face, use the cloth to wipe around the dock and tail head.

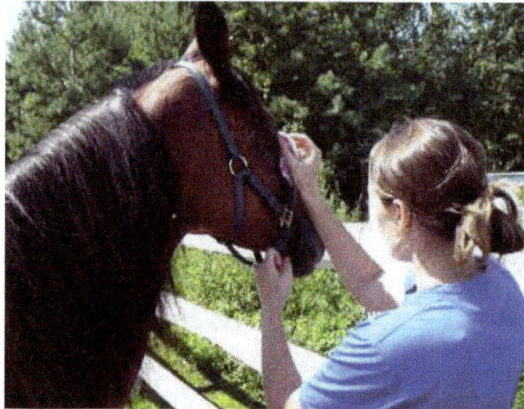

Gently clean around your horse's or pony's eyes, muzzle and dock area with a clean damp sponge or soft cloth

Apply Finishing Touches

Using hoof oil or ointment is a bit controversial. At best it does nothing beyond making the hooves look nice. Some people feel it prevents the hoof from absorbing necessary moisture; some feel it seals in moisture. Talk to your farrier to help you decide. Apply fly spray or sunscreen if conditions require. Some horses object to the hissing sound of sprayers, so go slowly and desensitize your horse gradually, perhaps using a bottle of water to start.

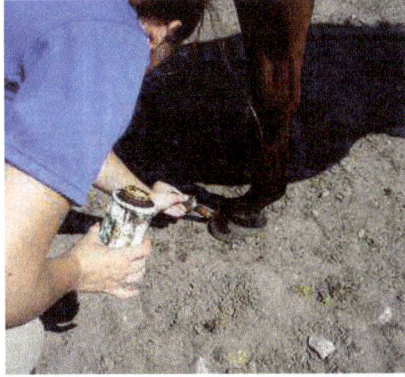

Apply hoof ointment to your horse's or pony's hooves to maintain moisture and prevent cracking if your farrier recommends it

Horse Sheath Cleaning

The sheath is an enfolded pocket of skin on the underside of the belly, just in front of the hind legs, that provides protection and support for the penis. When the penis is dropped for urination or breeding, most of the lining of the sheath stretches out and becomes visible along the penile shaft, but when the penis is retracted, all of that skin becomes bunched up on the inside of the sheath, much like a rumpled sleeve when you pull your arm backwards out of a shirt.

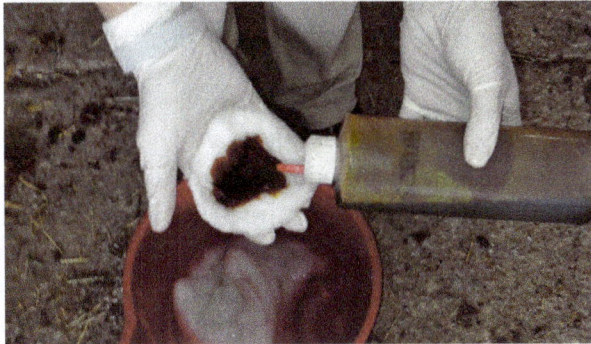

This bunched up and folded skin secretes natural oils and sheds skin, which combines with sweat and dirt to form a sticky, foul smelling residue called smegma. If you run your finger just inside the opening of the sheath, you will feel greasy clumps of dirt trapped in these loose folds, and if you look at the penis when it is dropped, you can see these ridges of dirt and sheets of flaking skin along the shaft.

At the very tip of the penis lies the opening of the urethra and a structure called the urethral fossa. Urine flows from the bladder through the urethra and exits through this small aperture. The fossa is a blind-ended pouch that lies adjacent to the urethral opening, forming a natural trap for skin, oil and debris. This accumulation is referred to as a "bean" and can become large enough to obstruct the flow of urine and be quite painful for the horse.

Studies have demonstrated that accumulated smegma can predispose a gelding to an aggressive type of skin tumour called squamous cell carcinoma on the penis, and regular cleaning to reduce this build-up can have a protective effect against this invasive cancer. Squamous cell carcinoma causes red, irritated, ulcerated patches on the skin and sometimes will have a wart-like appearance. They often bleed easily and are tender to the touch. Early recognition of this problem is key to effective management, which can involve freezing the masses with liquid nitrogen, topical chemotherapeutic agents, oral medication, and surgical resection of the diseased skin or of the entire end of the penis. Left untreated, this type of tumour can be quite invasive and extend into local lymph nodes so it is well worth your time to check this area regularly.

Time Gap in Sheath Cleaning

Every horse is different. Some geldings are naturally very dirty and benefit from cleaning every few months, while others are fine with a thorough cleaning once a year and the occasional "polish." Horses with unpigmented skin on the penis are more prone to squamous cell carcinoma and should at least be visually checked every couple of months, which you can easily do when the horse drops to urinate. Try to avoid too-frequent cleaning, as this can create a bacterial imbalance and allow more pathogenic species of bacteria and fungus to accumulate. In rare occasions, a chronic infection can occur, leading to a foul smelling discharge. In these cases, a skin swab should be submitted for culture and appropriate medical therapy instituted.

Things Needed for Cleaning the a Sheath

Sheath cleaning doesn't require special equipment. Any mild skin or dish soap, such as Ivory, or a commercial preparation from the tack shop is appropriate, but no matter what you choose, you must rinse thoroughly because residues in this area are very irritating. For this reason I would avoid any "leave in" treatments because a painful blister can occur in sensitive horses. You can accomplish the task very effectively with the following supplies:

- Mild soap
- Dish or exam gloves (these are essential, or the odour will cling to your hands and it is nearly impossible to get this dirt out from under your nails)
- Large (60cc) oral dose syringe or turkey baster
- Disposable sponges or roll cotton

- Clean warm water in two buckets

- Sedative/relaxant if required (only under veterinary supervision)

Safety First

Many geldings are comfortable having their sheaths cleaned and tolerate gentle traction on the penis in order to exteriorize it for a thorough examination. Others are a little tense and can benefit from medication to reduce tension on the penile retractor muscles. This is why one of the best times to clean a sheath is right after the vet has sedated them for another procedure such as dental work. Some horses are just too shy and ticklish in this area to allow you get this job done safely, and in those situations you are better off asking your vet to do the cleaning.

With time and patience, many can become accustomed to being handled in this area, but never get into a wrestling match trying to extract the penis from the sheath because you can damage the muscles and nerve supply, not to mention the danger of being kicked by an angry or frightened horse! Remember that even with the quietest and best behaved horse you are in a very vulnerable position with your head near the hind feet. Always err on the side of safety.

The Signs that the Horse Needs his Sheath Cleaned

The most obvious sign is shreds of dirt and skin on the penis when it is dropped to urinate. If the "bean" at the tip of the urethra is very large, the horse may actually have trouble urinating or the act will appear painful. A foul smell or cheesy looking discharge is a clear indication it is time to clean, and both mares and geldings may start to tail rub if there is significant dirt accumulation in the sheath, udder and beneath the tail. Even if you don't notice any of these obvious signs, it's still a good idea to clean and inspect the sheath a few times a year. It's a good idea to avoid sheath cleaning in extremely cold weather, especially of your horse needs a light tranquilizer to get the job done, which tends to leave them dropped and dangling for a while. Wet exposed flesh can freeze very quickly and significant frostbite damage can occur.

How to Clean a Sheath

Step 1. Using the sponge or cotton, dip it in the warm water and then insert your hand into the sheath and start to wet the area. Use only one of your buckets, and keep the second one clean for your rinsing. If your horse accepts it, you can use the turkey baster/syringe to squirt water into the sheath.

Step 2. Ideally, your horse will start to relax and his penis will drop down. Other more nervous horses may need some sedation to achieve this, or you can try to work "blind" (reaching inside the sheath), which is not ideal, but still moderately effective for most situations.

Step 3. Use a small amount of soap to help lubricate and loosen the greasy smegma. Remember that all the soap that goes in has to come out in the rinse, so start with a conservative amount.

Step 4. Work gently to clear the surface dirt and dead skin from the sheath and shaft of the penis. Sometimes the crusty material on the penis is still a bit adhered, so be careful when you pull this off.

Step 5. Time to extract the bean. Using a soapy finger (clipped nails please!), gently work it into the pocket that lies just above the urethral opening. If your horse tries to withdraw his penis, you can apply gentle traction on the shaft, but never get into a wrestling match because you can damage the retractor muscles.

Step 6. Once the bean is extracted, use your clean water and syringe or clean cotton/sponge to thoroughly rinse the entire area.

Hoof Care

Caring for your horse's feet and hooves will safeguard their long-term soundness. Following are some tips to keep your horse's feet healthy. Remember, no hooves, no horse.

Shoeing or Trimming Interval

Summer: In the summer, horses should be trimmed or shod at least every six to eight weeks. Performance horses may need more frequent trimming.

Winter: Hooves generally grow more slowly in the winter. Because of the slower growth horses should be trimmed every six to twelve weeks. The trimming or shoeing interval depends on each horse, and the amount of hoof they grow.

Hoof Balance

A balanced hoof allows the horse to move better, and puts less stress and strain on bones, tendons, and ligaments. e ideal foot has the following characteristics: a straight hoof-pastern angle, easy break over, adequate heel support, and mediallateral balance.

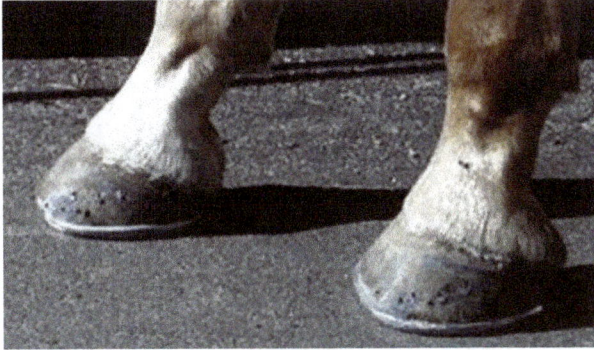

A balanced hoof

Straight hoof-pastern angle: ere is a straight line from the pastern down through the front of the hoof wall. is allows the bones to be aligned properly from pastern to coffin bone.

Medial-lateral balance: Thee foot lands evenly from side to side as the horse walks.

Easy break over: e toe is not too long and is squared, rounded or rolled. is allows easier movement with each step. Too much break over can result in health problems as well.

Adequate heel support: The shoe extends back to the end of the hoof wall and supports the back of the entire leg. Ideally, the back edge of the shoe is under a line drawn down the center of the cannon bone.

Hoof Wall Care

Weather conditions can cause damage to the hoof. During dry weather, or with frequent changes from wet to dry, horses are prone to having dry, brittle feet that easily develop hoof cracks. Prolonged trimming intervals can cause long toes, and the hoof wall often develops cracks due to the unsupported hoof wall. Unfortunately, some horses are born with poor hoof quality and are more susceptible to problems.

Treatment Tips: Apply hoof moisturizers to the hoof wall and sole during dry weather or if the hoof is brittle or developing cracks, proper nutrition and commercially available hoof supplements can help improve hoof quality, and most importantly, trim your horse on a regular basis.

Winter Hoof Care

In the winter, special care should be taken if your horse lives outside or is turned out. Snow can ball up under the sole and cause bruising or imbalance. Ice can be very slippery if the horse is shod with normal shoes.

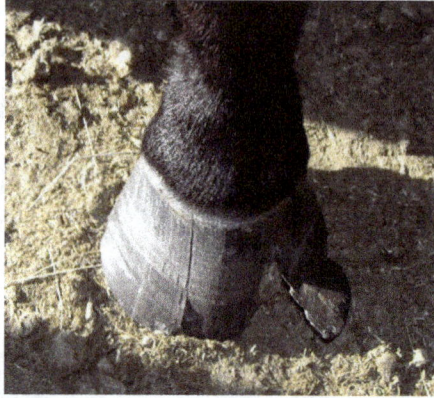

Hoof crack caused by long trimming interval

Winter tips: If your horse is normally barefoot, leave the shoes off. Horses usually slip less when barefoot or not shod. Horses that are prone to sole bruising may need shoes. If your horse is shod through the winter, have snow pads placed under the shoes and small cogs, borium, or nails placed at the heels. Snow pads will prevent snow and ice from building up under the shoe and the cogs or nails will allow for better traction. Finally, winter weather can dry out the hoof wall, and applications of hoof moisturizer may be needed.

Nutrition

Maintaining your horse's nutrition can help alleviate some hoof problems. Feeding good quality hay, supplementing the appropriate amount of vitamins and trace minerals, and making sure your horse has constant access to fresh, clean water is important for hoof health and overall horse health. Poor nutrition can lead to future hoof problems, and correcting a horse's nutrition can gradually improve hoof health over time. Cooperation between horse owners, veterinarians, and equine nutritionists are needed to ensure proper horse nutrition.

Research has shown that horses with poor quality hooves can benefit from commercially available hoof care products that contain Biotin (20 mg/day), Iodine (1 mg/day), Methionine (2500 mg/day) and Zinc (175 to 250 mg/day).

Common Hoof Problems

Poor shoeing or trimming: Long toes can results in collapsed heels, strain on flexor tendons and the navicular bone (Figure 3). If the horse is "too upright" it can cause trauma to the coffin bone and joint. An imbalanced hoof can cause stress on the collateral ligaments and joints.

Hoof cracks: Horizontal cracks or blowouts are usually caused by an injury to the coronary band or a blow to the hoof wall. Horizontal cracks or blowouts do not usually case lameness. Grass cracks are usually seen in long, unshod horses, and can be corrected with trimming and shoeing. Sand cracks results from injury to the coronary band or white line disease that breaks out at the coronary band. Sand cracks can be a cause of lameness. Treatment for sand cracks includes determining the cause and removing it, floating the hoof wall (not letting it bear weight), and/or fixation or patching of the crack. It usually takes nine to twelve months for the hoof to grow out.

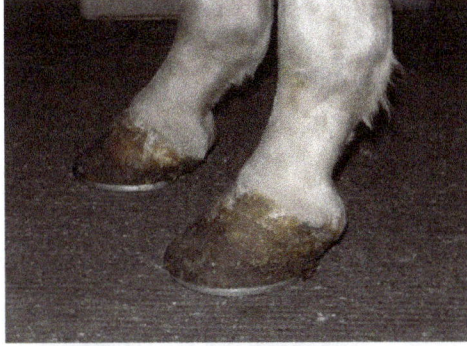

Long toes

Thrush: Thrush is a foul-smelling black exudate usually found around the frog that is associated with wet, soiled conditions. Thrush can invade sensitive tissue and cause lameness. Treatment includes keeping stalls or barn clean and dry can help eliminate thrush.

Solar abscess: Solar Abscess is an infection in the sole of the hoof that can lead to acute or severe lameness. Solar Abscess can be caused by trauma, bruising, or a foreign body. Treatments include removal of the foreign body (if possible), soaking the hoof in warm water and Epsom salt, and keeping the hoof bandaged, clean and dry.

Hot nail or street nail: A hot nail is a horseshoe nail that is driven into the sensitive structures of the hoof wall. Hot nails will usually cause lameness. Treatments include flushing the nail hole with antiseptic, packing the hole or bandaging the foot, and administering a Tetanus booster. A street nail is any foreign object that enters the foot. This is an emergency, and your veterinarian should be called immediately. Treatment depends on which hoof structure is affected.

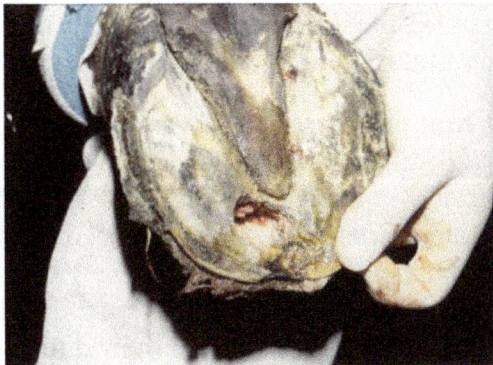

Solar abscess

Laminitis: Laminitis is inflammation of the sensitive laminae. Also called founder, laminitis is rotation (coffin bone rotates downward inside hoof capsule) and/or sinking (coffin bone sinks downward) of the coffin bone. There are several causes of laminitis. Treatments include regular shoeing or trimming, maintaining short toes, and frog and sole support.

Navicular: Navicular syndrome includes disease processes involving the navicular bone, bursa, ligamentous, and/or soft tissue structures. Horses will usually land their toe first due to pain in the heels. Causes of navicular syndrome include hereditary predisposition (Quarter Horses and Thoroughbreds), faulty conformation, hoof imbalance, and exercise on hard surfaces. Treatments include shoeing, maintaining a short toe, elevating the heels and good break over, and pads.

Natural Hoof Care

Booted horse on a trail ride. (Horse is in a transition period where
it cannot be ridden barefoot after shoe removal.)

Natural hoof care is the practice of keeping horses so that their hooves are worn down naturally and so do not suffer overgrowth, splitting and other disorders. Horseshoes are not used but domesticated horses may still require trimming, exercise and other measures to maintain a natural shape and degree of wear.

Within the natural hoof care philosophy, the term barefoot horses refers to horses which are kept barefoot full-time, as opposed to horses who are fitted with horseshoes. The hooves of barefoot horses are trimmed with special consideration to a barefoot lifestyle. The barefoot horse movement advocates a generalized use of barefoot horses, both in non-competitive and competitive riding, often coupled with a more natural approach to horse care. Horses are kept barefoot in many parts of the world, including South America, Mongolia and other industrialized and non-industrialized cultures.

Benefits of Barefooting

While horses have been used without shoes throughout history, the benefits of keeping horses barefoot has recently enjoyed increased popularity. Not only does the horse benefit with a healthier hoof in some cases, it can be less expensive to keep a horse barefoot, and many owners have learned to trim their horses' hooves themselves. As the health and movement benefits of barefooting have become more apparent in horses that have completed transition, horses are being competed barefoot in various sports (including dressage, show jumping, flat racing, steeplechase racing, trail riding and endurance riding).

Barefoot Trim

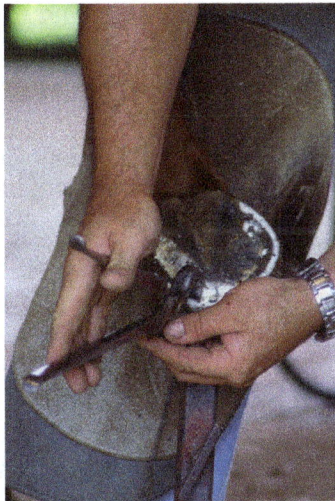

Hoof nippers are used to begin a trim of the hoof wall

There are several styles of barefoot trim in use today, including the Wild Horse or "Natural Trim" (developed by Jaime Jackson) the 4-Point Trim (Dr. Rick Reddin of NANRIC), the Strasser Trim (one of the most controversial as the horse's sole and bars are scooped out to widen the frog), the "Pete Ramey" trim where elements of the wild horse trim are the goal but the process includes removing hoof wall and forcing the horse to walk primarily on the sole. Some types, such as the 4-Point Trim can be used alone, or with shoes.

Barefoot trims are marketed to the public as something different from the "pasture" or "field" trim which farriers are trained to provide, taking into consideration hoof health and bony column angles, though each branded type of barefoot trim has its individual differences and there is no standardization or agreement between various barefoot advocacy groups. In contrast to farrier trims, barefoot trims are marketed as an approach to high performance hooves without the need for shoes, or simply as a natural approach to hoof care (depending upon the individual trimming method). However, they are something different, designed by nature itself to maintain a healthy, sound hoof without the use of shoes.

The barefoot trim aims to emulate the way in which hooves are maintained naturally in wild horse herds, like feral horse herds such as the American Mustang or the Australian Brumby, as well as wild zebras and other wild equine populations. Wild horses have been observed by Gene Ovnicek as having a hoof that tends to make contact with the ground on four points, and the hoof wall does not contact the ground at all. But the wild horse studies and measurements gathered by Jaime Jackson, a farrier at the time and working in unison with farrier Leslie Emery (author, Horseshoeing Theory & Practice) from 1982 to 1986 dispute Ovnicek's findings (The Natural Horse: Lessons from the Wild, 1992/1988 American Farriers Association annual conference). The trim guidelines he created for the AANHCP require the hoof wall to be on the ground as the most distal structure - with the sole, frogs and bars also acting as support structures when the horse is on uneven terrain. This is said to be another difference between the barefoot trim and the pasture trim, where the hoof wall was left long and in contact with the ground. Like wild horse populations, barefoot domestic horses can develop callouses on the soles of the hooves, allowing them to travel over all types of terrain without discomfort.

Important to the success of the barefoot trim is consideration for the domestic horse's environment and use, and the effects these have on hoof balance, shape, and the comfort of the horse. Objectives depend upon which method is followed: 1) many other than the AANHCP suggest shortening the hoof wall and heel to the outer edge of the concave sole for best hoof conformation, and 2) applying a rounded bevel ("mustang roll") to the bottom edge of the wall to allow for a correct breakover (the moment when the foot unloads and tips forward as it begins to lift off the ground) and to prevent chipping and flaring of the wall.

There is some research, but no scientific double blind studies, which indicates that removing horseshoes and using barefoot trimming techniques can reduce or in some cases eliminate founder (laminitis) in horses and navicular syndrome.

It is generally agreed upon by most natural hoof care practitioners that the management of the animal (diet and boarding conditions) are the most important components for the success of the horse to be barefoot. If the diet is unnatural, there will be inflammation and the horse cannot be comfortable.

Impact of Horseshoes

Removable iron horseshoes known as "hipposandals" may have been invented by the Roman legions. Nailed-on shoes were certainly used in Europe by the Middle Ages.

Horses were shod with nailed-on horseshoes from the Middle Ages to the present, though well-trained farriers also performed barefoot trimming for horses that did not require the additional protection of shoes. It has become standard practice to shoe most horses in active competition or work. However, there is a growing movement to eliminate shoes on working horses. Advocates of barefooting point out many benefits to keeping horses barefoot and present studies showing that improper shoeing can cause or exacerbate certain hoof ailments in the horse.

A hoof boot may help protect the horse's hooves during the transition period

Damage from improperly fitted and applied horseshoes can be seen in a gradual distortion of hoof shape, along with other ailments. Hoof soles are often sensitive when going barefoot after a long period of having been shod (because they are not thick enough through callusing). It can take weeks, months, a year, or more, depending on the horse's prior condition, before a horse is sound and usable on bare feet. During this transition period, the horse can be fitted with hoof boots which protect the soles of the feet until the horse has time to heal and build up callouses, though these boots, especially when not properly fitted and used, can cause hoof damage as well.

Hoof Health

The two things which can directly affect the health of the hoof are diet and exercise. Observers of wild horse populations note that the equine hoof stays in notably better condition when horses are in a herd situation and are free to move around 24 hours a day, as wild horses do, permitting good circulation inside the hoof. It is recommended that horses be allowed to walk at least 5 miles per day for optimum hoof health. The terrain should be varied, including gravel or hard surfaces and a water feature where the hooves can be wet occasionally.

Diet is very important too, as changes in feed can directly affect hoof health, most notably seen in cases of laminitis. Even some lots of hay may be high enough in sugar to cause laminitis. A healthy diet for horses currently with or prone to laminitis is based on free access to hay that has been tested for carbohydrate content and found to be less than 10% WSC + starch, some mineral supplementation, and no grain. Feeds and forage with high levels of sugar (carbohydrates) correlate with higher risk of clinical or subclinical laminitis and with other hoof ailments.

Natural hoof supplements can be used as a boost to the immune systems of horses when concerned with laminitis or other hoof ailments. D-Biotin supplements, often including the sulfur-containing amino acid dl-Methionine, are commonly known natural supplements that are effective for managing hoof health.

Modern research by individuals such as Jaime Jackson and Tia Nelson have studied feral horses to observe the way in which their natural foraging and roaming affects their hooves. They noticed that the hooves of these horses have a different configuration from domestic horses kept in soft pasture, having shorter toes and thicker, stronger hoof walls.

Controversies

Whether wearing shoes or going barefoot is better for the horse is the subject of some controversy. Opponents of the barefoot movement argue that domesticated horses are routinely put through abnormal levels of activity, stress, and strain, and their hooves undergo excessive wear and shock. Stable-kept horses are not exposed to the same environment as wild horses, which can affect their hoof quality. Additionally, humans sometimes favor certain traits over hoof quality (such as speed), and will breed horses with poor hoof quality if they are exceptional athletes. This can lead to overall decreased hoof quality within a breed and in riding horses in general. Advocates of traditional hoof care suggest that shoeing is needed to protect the hoof from unnatural destruction, and that the horseshoe and its various incarnations has been necessary to maintain the horse's usability under extreme and unnatural conditions.

Stable Bandage

A stable bandage, or standing bandage/wrap, is a type of wrap used on the lower legs of a horse. A stable bandage runs from just below the knee or hock, to the bottom of the fetlock joint, and protects the cannon bone, tendons of the lower leg, and fetlock joint.

A horse wearing standing bandages

Uses of the Stable Bandage

- Protection: the stable bandage offers some protection against minor cuts and bruises in a stall or horse trailer.

- Securing a poultice/dressing: stable bandages are often used to hold a poultice on the lower legs, or to hold on a wound dressing on an injury.

- To keep an injury clean: if a horse cuts his lower leg, a stable bandage can help keep the area from being contaminated by stall bedding or dirt. However, it may slow the healing process.

- Reduce or prevent "filling": after hard work, or if a horse is kept in a stall for long periods of time, the lower legs of the animal may "fill" or "stock up", causing filled legs (fluid builds up and swells the leg). A stable bandage can help prevent this.

- As a base: stable bandages are used as a "base" for bandages higher up on the leg (such as a knee or hock bandage). This prevents the swelling of the injury higher up from traveling down the leg.

- When bandaging in pairs: when a horse injures a leg, it often places more weight, and thus excess stress, on the uninjured leg. To prevent the uninjured leg from swelling, it should also be bandaged. So both front legs, both hind legs, or all four legs should be bandaged.

- Traveling: stable bandages are often used when shipping a horse, instead of using shipping bandages (which are more time-consuming to apply), or shipping boots (which may not offer as much protection). When used for shipping, it is best to also use bell boots on the front legs, as the heels and pasterns are not protected by a stable bandage.

Dangers of the Stable Bandage

An incorrectly applied stable bandage may do more harm than good. Therefore, it is important to learn from an experienced horseman, and to practice bandaging before keeping bandages on for long periods of time. Considerations to be aware of when bandaging include:

- Tightness of the wrapping: If the wrapping is not tight enough, the bandage may slip down and possibly trip the horse. If it is too tight, or uneven, it may cut off circulation to the lower leg or cause "cording" or bandage bows.

- Padding above and below the wrap: If too much padding is left above or below the bandage material, it may catch on something, and dislodge the bandage or frighten the caught horse.

- Ending the bandage: The closures of bandage materials may catch on each other, or on the bandage of the opposite leg, if the bandage closures end on the inside of the leg. This could dislodge or open the bandage. Therefore, bandages should always end on the outside of the horse's leg.

Using a Bandage with a Poultice or Leg Dressing

Bandages should typically be changed every 12 hours. When a poultice is used, it is best to clean the leg with soap and water after removal, before re-applying a clean bandage.

Before a wound dressing is applied, the injury should be cleaned, and it should be re-cleaned between bandage cycles. The wrapping of the padding and bandage should be in the direction that best supports the closure of the wound. Gauze should be used between the padding and the leg injury.

Horse Blanket

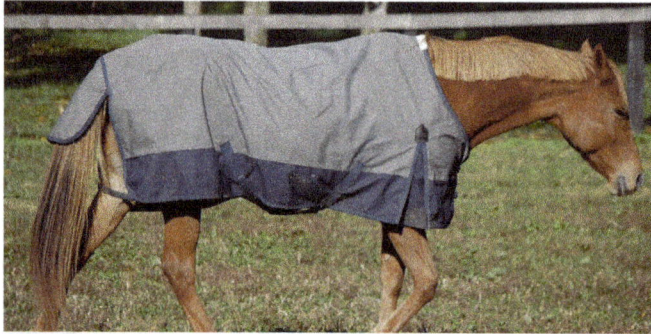

A horse blanket or rug is a blanket or animal coat intended for keeping a horse or other equine warm or otherwise protected from wind or other elements. They are tailored to fit around a horse's body from chest to rump, with straps crossing underneath the belly to secure the blanket yet allowing the horse to move about freely. Most have one or two straps that buckle in front, but a few designs have a closed front and must be slipped over a horse's head. Some designs also have small straps that loop lightly around the horse's hind legs to prevent the blanket from slipping sideways.

Protection from the Elements

Standard horse blankets are commonly kept on a horse when it is loose in a stall or pasture as well as when traveling. Different weights are made for different weather conditions, and some are water-resistant or waterproof. Modern materials similar to those used in human outdoor wear are commonly used in blanket manufacture.

Blankets are sometimes used to keep the horse's hair short. If horses are blanketed at the beginning of the autumn, especially if kept in a lighted area for 16 hours a day, they will not grow a winter coat. Blankets also protect horses that are kept with a short clipped hair coat for show purposes. When a horse is given a full body clip, or even a partial "trace clip", it needs to have a blanket kept on at all times if the weather is cool because the horse no longer has the natural insulation of a longer hair coat. If a blanket is put on a horse at the beginning of the winter in order to suppress the growth of a winter coat, or if the horse is kept clipped in cold weather, the blanket cannot be taken off until warmer weather arrives in the spring. If a horse is subjected to cold weather without either a blanket or a natural hair coat to keep it warm, it may become ill, and vulnerable to sicknesses such as influenza.

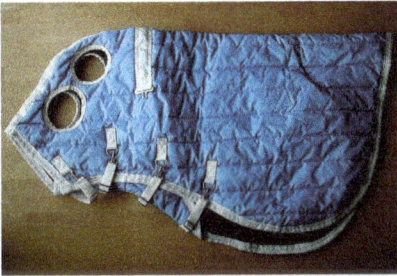

A hood, showing openings for eyes and ears

Horses wearing knotted cord fly sheets

Heavy blankets for warmth make up the bulk of the horse blanket market, but lightweight blankets may be used in the summer to help the animal ward off flies and to prevent the hair coat from bleaching out. Such blankets are usually called a "sheet" or a "fly sheet". They are usually made of some type of nylon or strong synthetic fiber, but with the capacity to "breathe" so that the animal remains cool. Most have a smooth nylon lining in front to prevent hair from wearing off on the shoulders. They are becoming increasingly popular, particularly with the rise of insect borne diseases such as West Nile Virus.

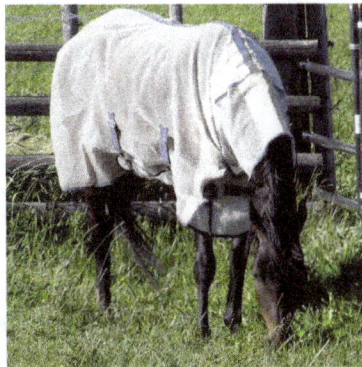

A horse wearing a summer fly sheet with attached neck cover, light blankets ward off insects and prevent coat bleaching

Any blanket may be supplemented with a neck cover or a full hood. Neck covers are often attached directly to the blanket. Hoods are a separate piece of horse "clothing", which cover the neck and come down the head to just above the muzzle of the horse, with holes cut for the eyes and ears. Summer weight hoods and neck covers help keep away insects and are also frequently used to keep a horse clean before a horse show. Winter weight hoods are used for warmth.

Saddle Blankets

A police horse wearing a quarter sheet

A blanket or pad used under a saddle when a horse is being ridden is called by many names, including a saddle blanket, saddle cloth, numnah, and saddle pad. They usually do not cover the horse's entire body, though a hybrid design that is a cross between a saddle blanket and a horse blanket, called a quarter sheet, is a blanket placed under the saddle but which covers the horse from shoulder to hip while riding. Quarter sheets are sometimes used in cold weather to keep a horse's muscles loosened up when warming up for competition, or on horses that may have to stand around when under saddle and run the risk of stiffening up if their muscles get chilled.

Other Designs

A horse wearing a cooler

A cooler or a mantle, is a large, nearly square blanket with ties that is draped over a horse that is hot and sweaty from an intense workout, or one that has just been bathed and is wet all over. It is commonly made of wool or synthetic fleece, though a few designs are made of woven cotton. It is worn as the horse is being walked to cool down and allows enough air circulation for the horse to dry, but slows the rate of drying to prevent the horse from becoming hypothermic. It is designed so it can be tied shut in front; most designs have a small browband which can be used to keep it positioned well up on the neck, and it may have a loose cord that goes beneath the tail to prevent the wind from blowing it off from the rear, but usually it has no other straps or attachments. It is intended to be used on a horse while the animal is being led or held by a person, and not left on an unattended horse. In windy weather, a loose surcingle may be added to prevent the cooler from blowing completely off.

A traditionally-shaped blanket of loosely crocheted cotton, called an anti-sweat sheet, is used to absorb moisture. Often used alone to wick moisture from the surface of the horse, if placed under a cooler, it is removed when it becomes wet.

A barrier blanket is sometimes used in Australian Thoroughbred racing. This blanket weighs about 40 kilograms and is placed over the horse prior to entering the starting stalls. It is then tied to the back of the stalls after the horse has been loaded and remains in the barrier as the horse jumps. Quite a few horses respond positively to it and are more easily loaded and are calmer in the stalls.

References

- Preparation for Foaling by Brad Dowling BVSc MVetClinStud FACVSc Archived 2011-02-07 at the Wayback Machine Retrieved 2011-2-7

- Juan C. Samper (1 January 2009). Equine Breeding Management and Artificial Insemination. Elsevier Health Sciences. ISBN 1-4160-5234-8

- Hura, V; et al. (October 1997). "The effect of some factors on gestation length in nonius breed mares in Slovakia (Egyes tényezők hatása a nóniusz fajta vemhességének időtartamára)". Proceedings of Roundtable Conference on Animal Biotechnology. XIII. Retrieved 2008-04-22

- Stratton, Charles, The International Horseman's Dictionary, Lansdowne Press, Melbourne, 1978, ISBN 0-7018-0590-0

- Lewis, Barbara S. "Egyptian Arabians: The Mystique Unfolded". Arabians. Pyramid Arabians. Archived from the original on 2006-05-08. Retrieved 2006-05-10

Chapter 6

Various other Types of Livestock Farming

A wide range of species are reared as livestock. Some of the common livestock animals are sheep, goat, pig, etc. This chapter elucidates the different techniques and methods associated with the farming of the aforementioned livestock animals and their advantages. It has been carefully written to provide an easy understanding of the varied aspects of livestock farming.

Sheep Farming

Raising sheep can be a fun and profitable venture but, before you begin, be sure you know what is involved. Most farmers raise sheep for their meat and/or wool. Some choose to raise sheep for their milk. What are your intentions? Will this be an additional revenue stream, a family food source, or merely a hobby? At Southern States, we know that any farming opportunity, including raising sheep, requires proper planning and the right equipment to be successful. Here is a brief overview of a sheep farmer's requirements and responsibilities.

Raising sheep can be a great option for farmers and others who don't have enough land to keep cattle but still want to raise some type of livestock. The general consensus amongst sheep farmers is that you can have 4 to 6 sheep per acre as opposed to one cow and her calf per acre. Sheep are social creatures and are happiest in a group setting, so it is recommended that you keep a minimum of four or five sheep. Once mature, sheep can range from 50 lbs to 500 lbs depending on the breed.

Before starting any new pursuit, it is important to understand the costs involved. Fortunately, sheep can be relatively inexpensive to raise. Start-up costs include the price of the sheep, fencing, and shelter. You may even have the shelter and fencing already set up. On-going costs include any sheep feed required and health costs, including vaccinations and vet visits. Raising sheep does not require a lot of expensive machinery. A mower may be needed to cut overgrown fields, but proper pasture management should limit the need for this. Basic supplies will include a set of clippers for shearing and hoof trimmers.

Sheep do not require complex or expensive shelter. They can survive most of the time outside provided there is some basic shelter from the wind and bad weather. Their housing can vary depending on your budget and your climate. The ideal accommodations are dry, open-sided, and well ventilated. Hay, straw, or pine shavings used as bedding can provide extra warmth and comfort for your sheep.

Feeding your sheep can be quite affordable. Keep a fresh supply of water accessible at all times. Pasture grass is the main source of food for the sheep herd, but sometimes you may have to supplement with sheep feed and other grains. Proper grazing management will not only reduce costs associated with supplemental feed, but it will also help manage overgrowth. Develop a timetable for rotating your sheep throughout your pastures because if left too long in one location, sheep can wreak havoc on your fields. Overgrazing of pastures can kill the grasses, lead to soil erosion and increase farm costs associated with re-seeding.

While sheep are less susceptible to diseases than other livestock breeds, they're more prone to parasites and extremely vulnerable to predators. Fortunately, the sheep owner can help ensure his or her flock's health and safety. Be diligent about preventing internal parasite infestation which can cause death in young lambs. Predators are a serious danger to sheep and include animals such as coyotes, wolves, mountain lions and dogs. A sturdy, well-built fence will help keep predators out. Guard animals can also protect your sheep. According to the U.S. Department of Agriculture's Wildlife Services, some recognized breeds used for livestock protection dogs include: Great Pyrenees, Anatolian Shepherds, Maremmas and Komondors.

Sheep Dip

Sheep dip is a liquid formulation of insecticide and fungicide which shepherds and farmers use to protect their sheep from infestation against external parasites such as itch mite (Psoroptes ovis), blow-fly, ticks and lice.

Plunge dipping sheep

Design

Sheep dip is available as wettable powders, pastes, solutions, or suspensions which are used to prepare diluted solutions or suspensions. The term is used both for the formulation itself, and the trough in which the sheep is completely immersed.

There are two broad classes of sheep dip: organophosphorus compounds, from which chemical warfare agents were later developed, and synthetic pyrethroids. Organophosphorous compounds are very toxic to humans exposed even to very low doses, as they travel easily through the skin, and are cumulatively toxic.

Plunge sheep dips may be permanent in-ground structures, often of concrete, or mobile transportable steel structures. Invented after the permanent plunge dip was the rotating, power spray dip. These dips are redundant in the major sheep breeding countries, as the backliners and jetting provide a better alternative.

Negative Health and Environmental Effects

Sheep dips have been found to cause soil contamination and water pollution. They contain chemical insecticides that are highly toxic to aquatic plants and animals. For this reason, it is important that the dip and dipped sheep are well managed to avoid spreading the chemicals and causing water pollution.

Some chemicals used in sheep dips are known to have been harmful. A sheep dip based on organo-phosphates has resulted in neurological conditions known as "OP poisoning".

Advantages of Sheep Farming

Advantages of sheep farming:-

1. Multi-faceted utility: meat, wool, skin, manure, and to some extent milk & transport helps it to play an important role in the Indian agrarian economy.

2. The production of wool, meat and manure provides three different sources of income per year.

3. Since the two major products of sheep (wool and mutton) are entirely different in their production and utilization, the price of one may not necessarily have a bearing on the other. Wool may be stored and held for higher prices or sold at shearing time. A crop of lambs may be marketed from 5-6 months onwards (preferably before one year), bringing rather a quick return.

4. Mutton is one kind of meat towards which there is no prejudice by any community in India.

5. In addition to wool, mutton and to some extent milk, sheep provide employment to about 3 million people in the form of self-employment, as hired labour for tending flocks during migration, and persons engaged in wool shearing and in wool and skin processing. Furthermore, sheep farming is a logical source of livelihood in arid zones where crop production is an uncertainty and thus it suitably fits into desert development programmes in vogue by protecting them from the vagaries of drought and famine.

6. Most suitable of the small ruminants to utilize the sparse vegetation in dryland areas through rangeland management and developed (reseeded) pasture.

7. Unlike goats, sheep hardly damage any tree.

8. Better adapted to arid and semi-arid tropics with marginal and sub-marginal lands, otherwise unfit for crops, due to their superior water & feed (esp. protein) economy.

9. Since sheep eat more different type of plants than any other kind of livestock, they can turn waste into profit and at the same time improve the appearance of many farms (i.e. excellent weed destroyer).

10. Sheep dung is a valuable fertilizer, and since they are grazed on sub-marginal lands, their droppings are the only means of improving the growth of plants in such areas.

Unique characteristics of sheep:-

1. Strong herd instincts of sheep make them excellent ranch animals as they keep together in tight and easily managed flocks and do not disperse widely

all over the available land, which would make it difficult to protect them from predators and difficult to round up.

2. Excellent ability to survive over a prolonged period of drought and semi-starvation.

3. Sheep have the ability to produce prime carcasses on roughage alone, thus they are well adapted to many areas unable to produce grain profitably.

4. The structure of their lips helps them to clean grains lost at harvest time, and thus convert waste feed into profitable products.

5. Less prone to extreme weather conditions, ectoparasites as well as other diseases.

6. Unique ever-growing fibre which allows ventilation and also protects the skin from the hot sun, rain and abrasions.

7. Sheep can also constrict or relax blood vessels in the face, legs and ear for control of heat loss.

8. Their visual sense is exceedingly well-developed they can discern movement far better than humans, but cannot distinguish shapes as well as man.

9. Sheep do not need expensive buildings to house them.

10. Sheep require less labour than other kinds of livestock.

Because of their hardiness and adaptability to dry conditions, the north-western and southern peninsular regions of the country have a large concentration of sheep. In the tropics, they are non-seasonal breeders and can be made to lamb throughout the year.

Because of their close grazing nature and ability to utilize very low-set vegetation which no other animal can utilize and their capacity to cover long distances in search of forage and water, they have often been generally associated with desertification. In reality, it is not the sheep but the man who owns the sheep, who is to be blamed for the misconceived management of the grazing lands, leading to desertification. Rather, a controlled and judicious grazing on the non-cultivable land could prevent soil erosion and make it fertile and suitable for crop production through sheep dropping collected over the years.

There is great variation in the external characteristics of sheep, manifested in the number and form of horns in the shape and size of ears, in an arching of nasal bones in some types, in length of tail and in the development of great masses of fat at the base of the tail and other posterior parts of the body. There are extreme variations too in colour of the face and other parts not covered with wool. Great variations exist in the quality and colour of the fleece. These variations have provided the basis for improving sheep for different products viz. wool, mutton and pelt. Variations in wool were pronounced with respect to colour, staple length, fineness and other characteristics.

The wool on the shoulder is finer than that grows on the thigh, belly and around the tail. Wool that grows on the folds in the skin is likely to be considerably coarser than that which grows between the folds.

Goat Farming

Probably first domesticated in the East, perhaps during prehistoric times, the goat has long been used as a source of milk, cheese, mohair, and meat. Its skin has been valued as a source for leather. In China, Great Britain, Europe, and North America, the goat is primarily a milk producer. By good management its limited (six months per year) breeding season and the consequent difficulty of maintaining a level supply of milk throughout the year, can be overcome. The goat is especially adapted to small-scale production of milk for the family table; one or two goats supply sufficient milk for a family throughout the year and can be maintained economically in quarters where it would not be practical to keep a cow.

Pure-white goat's milk compares favourably with cow's milk in flavour and keeping qualities under sanitary conditions. It has certain characteristics differing from cow's milk that make it more easily digested by infants, invalids, and persons with allergies. Goat flesh is edible, that from young kids being quite tender and more delicate in flavour than lamb, which it resembles. Goat flesh is much prized in the Mediterranean countries, particularly in Spain, Italy, the south of France, and Greece. The Angora and Cashmere breeds are famous for their fine wool or mohair.

The many breeds may be roughly grouped: the prickeared—e.g., Swiss goats; the eastern, or Nubian, with long, drooping ears; and the wool goat—e.g., Angora. While it is usually easy to distinguish goats from sheep, certain hair breeds of the latter are, to the layman, only distinguishable from goats by the direction of the tail, upward in goats, downward in sheep.

Of the Swiss goats, from which many of the best modern breeds are derived, the Toggenburg and Saanen are most important. The French breeds have much Swiss blood. In Germany the many varieties trace to Swiss breeds, which are also popular throughout Scandinavia and the Netherlands.

The Maltese goat, an important source of milk on the island of Malta, probably contains eastern blood. Many goats are found in Spain, northern Africa, and Italy, among them the Murcian, Granada, and La Mancha.

Nubians are African goats, chiefly Egyptian. They are usually large, short-haired goats with large lop ears and Roman noses. They may be of solid colour, parti-coloured, or spotted. The goats in Israel and Syria have long hair and large lop ears and most commonly are solid black or with white spots. Most Indian varieties, the best of which come from the Yamuna River area, have lop ears.

In Britain, the native goat was small, with short legs, long hair—usually gray but of no fixed colour—and with no definite markings. The widespread use of pedigree males, mostly of Swiss extraction, to improve the milk yield, has resulted in the almost total disappearance of the native types.

The Boer goat is a widely farmed meat breed

Goat farming is the raising and breeding of domestic goats (Capra aegagrus hircus). It is a branch of animal husbandry. Goats are raised principally for their meat, milk, fibre and skin.

Goat farming can be very suited to production with other livestock such as sheep and cattle on low-quality grazing land. Goats efficiently convert sub-quality grazing matter that is less desirable for other livestock into quality lean meat. Furthermore, goats can be farmed with a relatively small area of pasture and limited resources.

Pasture

Goats are remarkably agile and will climb trees to browse

As with other herbivores, the number of animals that a goat farmer can raise and sustain is dependent on the quality of the pasture. However, since goats will eat vegetation that most other domesticated livestock decline, they will subsist even on very poor land. Therefore, goat herds remain an important asset in regions with sparse and low quality vegetation.

Meat

Three-quarters of the global population eat goat meat. It comprises 10% of worldwide meat consumption and 60% of red meat.

Goat meat contains low amounts of saturated fatty acids and cholesterol. It is considered to be a healthier alternative to other types of red meat.

The taste of goat kid meat has been reported as similar to that of spring lamb meat. In some localities (e.g. the Caribbean, Bangladesh, Pakistan and India) the word "mutton" is used to describe both goat and lamb meat. However, some compare the taste of goat meat to veal or venison, depending on the age and condition of the goat. The flavour is primarily linked to the presence of 4-methyloctanoic and 4-methylnonanoic acid.

Goat meat can be prepared in a variety of ways, including stewing, baking, grilling, barbecuing, canning, and frying; it can be minced, curried, or made into sausage. Because of its low fat content, the meat can toughen at high temperatures if cooked without additional moisture. One of the most popular goats farmed for meat is the South African Boer, introduced into the United States in the early 1990s. The New Zealand Kiko is also considered a meat breed, as is the myotonic or "fainting goat".

Offal

When slaughtered, none of the animal need be wasted – so, all of the eviscerate can be used. This includes heart, liver, kidneys, spleen, lungs, stomach and intestines.

The stomach can be eaten as tripe or stuffed like a haggis.

The intestines make good sausage casings but have to be processed whilst still warm, otherwise the digestive enzymes start to digest the gut itself post mortem. Once cleaned out and cut into convenient lengths they can be turned inside out so that the internal lining can be scraped away with a knife. They can then be left in an isotopic cool solution (salty water) which keeps them supple until needed. This is traditionally done in autumn/fall when the temperature drops to below 55 deg Fahrenheit to avoid the meat within rotting -which can lead to botulism poisoning. Likewise, the skins need to be pricked so that air can ingress, as botulism needs an anaerobic medium in order to multiply. Goat sausages dry cured in this way, require no refrigeration. Modern curing methods include nitre to inhibit botulism but it is best not to wean a baby on sausages as it could induce infantile botulism from any spores present (this is not something that an older child or adult should be concerned about because honey often contains botulism spores since the bacteria is ubiquitous in our environment but a neonate's gut has less of a immune response). It is considered by some that such dry cured sausages of goat, take on a rancid taste quite quickly and develop a strong taste so are not popular today. Yet, for those that like a goaty taste it only adds to their appeal.

In the case of billy goats, the testicles sliced and fried are considered a delicacy.

Milk, Butter and Cheese

Goats produce about 2% of the world's total annual milk supply. Some goats are bred specifically for milk. Unprocessed goat milk has small, well-emulsified fat globules, which means the cream remains suspended in the milk instead of rising to the top, as in unprocessed cow milk; therefore, it does not need to be homogenized. Indeed, if goat milk is to be used to make cheese, homogenization is not recommended, as this changes the structure of the milk, affecting the culture's ability to coagulate the milk and the final quality and yield of cheese. Dairy goats in their peak milk production (generally around the third or fourth lactation cycle) average—2.7 to 3.6 kg (6 to 8 lb)—of milk production daily—roughly 2.8 to 3.8 l (3 to 4 U.S. qt)—during a ten-month lactation, producing more just after freshening and gradually dropping in production toward the end of their lactation. The milk generally averages 3.5% butterfat.

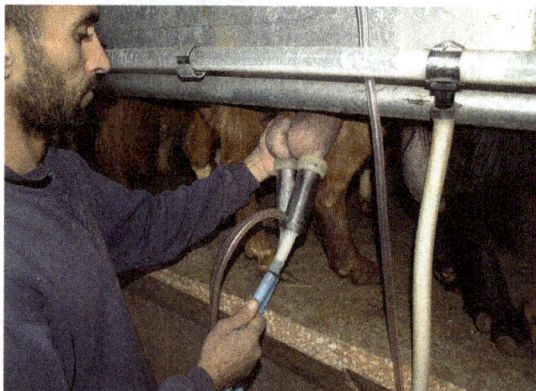

A goat being machine milked on an organic farm

Goat milk is commonly processed into cheese, butter, ice cream, yogurt, cajeta and other products. Goat cheese is known as fromage de chèvre ("goat cheese") in France. Some varieties include Rocamadour and Montrachet. Goat butter is white because goats produce milk with the yellow beta-carotene converted to a colourless form of vitamin A.

Male goats are generally not required for the dairy-goat industry and are usually slaughtered for meat soon after birth. In the UK, approximately 30,000 billy goats from the dairy industry are slaughtered each year.

Fibre

Most goats have soft insulating hairs near the skin and longer guard hairs on the surface. The desirable fibre for the textile industry is the former; it has several names including "down", "cashmere" and "pashmina". The guard hairs are of little value as they are too coarse, difficult to spin and difficult to dye. Goats are typically shorn twice a year, with an average yield of about 4.5 kg (10 lb).

An Angora goat

In South Asia, cashmere is called "pashmina" (from Persian pashmina, "fine wool"). In the 18th and early 19th centuries, Kashmir (then called Cashmere by the British), had a thriving industry producing shawls from goat-hair imported from Tibet and Tartary through Ladakh. The shawls were introduced into Western Europe when the General in Chief of the French campaign in Egypt (1799–1802) sent one to Paris. Since these shawls were produced in the upper Kashmir and Ladakh region, the wool came to be known as "cashmere".

The cashmere goat produces a commercial quantity of cashmere wool, which is one of the most expensive natural fibres commercially produced; cashmere is very fine and soft. The cashmere goat fibre is harvested once a year, yielding around 260 g (9 oz) of down.

Angora goats produce long, curling, lustrous locks of mohair. Their entire body is covered with mohair and there are no guard hairs. The locks constantly grow to 9 cm or more in length. Angora crossbreeds, such as the pygora and the nigora, have been selected to produce mohair and/or cashgora on a smaller, easier-to-manage animal.

Goat Skin

The skin of goats is a valuable by product of goat farming. Up until 1849 all Rolls of Parliament were written upon parchment usually made from goat skin. Another popular use is for drum skins. Parchment is prepared by liming (in a solution of quick lime) to loosen the hair follicles. After several days in this bath, the hair can then be scraped off and the under surface of the skin scraped clean. After that the finished skins are sewn into a wooden frame to dry and shrink.

Finished parchment made of goatskin stretched on a wooden frame

Parchment is still available today, but imported skins can carry a small risk of harboring anthrax unless properly treated.

Goat farm

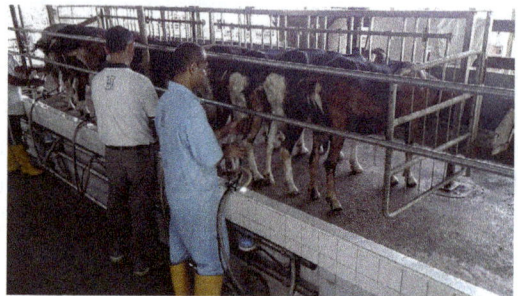

Goat milking

Pig Farming

Pig farming or pig husbandry, the raising of pigs for meat, may be organized in a more or less sophisticated way, but it can always be analyzed on the basis of the production cycle and is divided into two parts:

- farrowing sows for the production of weaned piglets;

- the rearing of those piglets, as future breeding animals or as pigs for slaughtering.

In a non-organized sector the genetic selection is managed within each herd, with the breeding animals being kept from the grown pigs (gilts) or purchased outside (boars) and used for natural service.

In a more sophisticated organisation the breeding animals are produced in specialized farms taking part in a thorough selection scheme (nucleus-multipliers-breeders). The sows are usually inseminated artificially and patch farrowed. Depending on the efficiency of the selection scheme, the pig fattening performance is improved. The functions of breeders, farrowers and fatteners can be combined and the fattening can be divided into growing and finishing.

The housing of pigs may vary from confined buildings for rapid standardized production to outdoor extensive production, as well as all the forms of organization in between. Most of the production is conducted according to optimized intensive methods, and extensive farming may either be in response to consumer demand (e.g. organic farming) or be used for own consumption. In such cases, the piglets grown and fattened in winter make use of cereals, food waste, by-products, buildings or labour force.

Pigs are amenable to many different styles of farming: intensive commercial units, commercial free range enterprises, or extensive farming (being allowed to wander around a village, town or city, or tethered in a simple shelter or kept in a pen outside the owner's house). Historically, farm pigs were kept in small numbers and were closely associated with the residence of the owner, or in the same village or town. They were valued as a source of meat, fat and for the ability to turn inedible food into meat, and often fed household food waste if kept on a homestead. Pigs have been farmed to dispose of municipal garbage on a large scale.

All these forms of pig farm are in use today. In developed nations, commercial farms house thousands of pigs in climate-controlled buildings. Pigs are a popular form of livestock, with more than one billion pigs butchered each year worldwide, 100 million of them in the USA. The majority of pigs are used for human food but also supply skin, fat and other materials for use as clothing, ingredients for processed foods, cosmetics, and medical use.

The activities on a pig farm depend on the husbandry style of the farmer, and range from very little intervention (as when pigs are allowed to roam villages or towns and dispose of garbage) to intensive systems where the pigs are contained in a building for the majority of their lives. Each pig farm will tend to adapt to the local conditions and food supplies and fit their practices to their specific situation.

The following factors can influence the type of pig farms in any given region:

- Available food supply suitable for pigs

- The ability to deal with manure or other outputs from the pig operation

- Local beliefs or traditions, including religion

- The breed or type of pig available to the farm

- Local diseases or conditions that affect pig growth or fecundity

- Local requirements, including government zoning and/or land use laws

- Local and global market conditions and demand

Use as Food

Almost all of the pig can be used as food. Preparations of pig parts into specialties include: sausage (and casings made from the intestines), bacon, gammon, ham, skin into pork scratchings, feet into trotters, head into a meat jelly called head cheese (brawn), and consumption of the liver, chitterlings and blood (blood pudding or black pudding). This is also, technically, the case for all other mammals, although the demand isn't really there.

Production and Trade

Pigs are farmed in many countries, though the main consuming countries are in Asia, meaning there is a significant international and even intercontinental trade in live and slaughtered pigs. Despite having the world's largest herd, China is a net importer of pigs, and has been increasing its imports during its economic development. The largest exporters of pigs are the United States, the European Union, and Canada. As an example, more than half of Canadian production (22.8 million pigs) in 2008 was exported, going to 143 countries. Older pigs will consume eleven to nineteen litres (three to five US gallons) of water per day.

Relationship between Handlers and Pigs

The way in which a stockperson interacts with pigs affects animal welfare which in some circumstances can correlate with production measures. Many routine interactions can cause fear, which can result in stress and decreased production.

There are various methods of handling pigs which can be separated into those which lead to positive or negative reactions by the animals. These reactions are based on how the pigs interpret a handler's behavior.

Negative Interactions

Many negative interactions with pigs arise from stock-people dealing with large numbers of pigs. Because of this, many handlers can become complacent about animal welfare and fail to ensure positive interactions with pigs. Negative interactions include overly-heavy tactile interactions (slaps, punches, kicks and bites), the use of electric goads and fast movements. It can also include killing them. However, it is not a commonly held view that death is a negative interaction. These interactions can result in fear in the animals, which can develop into stress. Overly-heavy tactile interactions can cause increased basal cortisol levels (a "stress" hormone). Negative interactions that cause fear mean the escape reactions of the pigs can be extremely vigorous, thereby risking injury to both stock and handlers. Stress can result in immunosuppression, leading to an increased susceptibility to disease. Studies have shown that these negative handling techniques result in an overall reduction in growth rates of pigs.

Positive Interactions

Various interactions can be considered either positive or neutral. Neutral interactions are considered positive because, in conjunction with positive interactions, they contribute to an overall non-negative relationship between a stock-person and the stock. Pigs are often fearful of fast movements. When entering a pen, it is good practice for a stock-person to enter with slow and deliberate movements. These minimize fear and therefore reduce stress. Pigs are very curious animals. Allowing the pigs to approach and smell whilst patting or resting a hand on the pig's back are examples of positive behavior. Pigs also respond positively to verbal interaction. Minimizing fear of humans allow handlers to perform husbandry practices in a safer and more efficient manner. By reducing stress, stock are made more comfortable to feed when near handlers, resulting in increased productivity. In other words, pigs are very social and intelligent animals, and if they are treated well, better meat can be obtained. Prohand for pigs is a training program that teaches handlers to interact with pigs in a way that promotes safe handling. It promotes the development of positive behaviors and elimination of negative behaviors. This program has been seen to improve productivity without any capital investment.

Terminology

Pigs are extensively farmed, and therefore the terminology is well developed:

- Pig, hog or swine, the species as a whole, or any member of it. The singular of "swine" is the same as the plural.

- Shoat, piglet or (where the species is called "hog") pig, unweaned young pig, or any immature pig.

- Sucker, a pig between birth and weaning.

- Weaner, a young pig recently separated from the sow.

- Runt, an unusually small and weak piglet, often one in a litter.

- Boar or hog, male pig of breeding age.

- Barrow, male pig castrated before puberty.

- Stag, male pig castrated later in life (an older boar after castration).

- Gilt, young female not yet mated, or not yet farrowed, or after only one litter (depending on local usage).

- Sow, breeding female, or female after first or second litter.

Pigs for Slaughter

- Suckling pig, a piglet slaughtered for its tender meat.

- Feeder pig, a weaned gilt or barrow weighing between 18 kg (40 lb) and 37 kg (82 lb) at 6 to 8 weeks of age that is sold to be finished for slaughter.

- Porker, market pig between 30 kg (66 lb) and about 54 kg (119 lb) dressed weight.

- Baconer, a market pig between 65 kg (143 lb) and 80 kg (180 lb) dressed weight. The maximum weight can vary between processors.

- Grower, a pig between weaning and sale or transfer to the breeding herd, sold for slaughter or killed for rations.

- Finisher, a grower pig over 70 kg (150 lb) liveweight.

- Butcher hog, a pig of approximately 100 kg (220 lb), ready for the market. In some markets (Italy) the final weight of butcher pig is in the 180 kg (400 lb) range. They to have hind legs suitable to produce cured ham.

- Backfatter, cull breeding pig sold for meat; usually refers specifically to a cull sow, but is sometimes used in reference to boars.

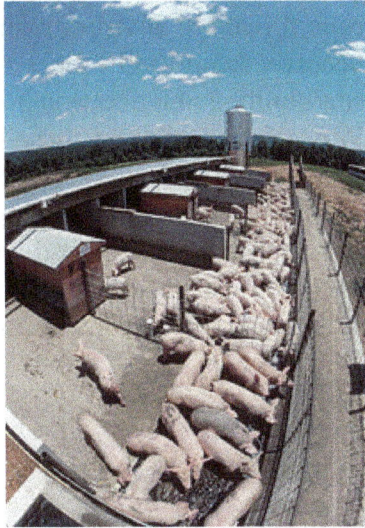

Finishing pigs on a farm.

Groups

- Herd, a group of pigs, or all the pigs on a farm or in a region.
- Sounder, a small group of pigs (or wild boar) foraging in woodland.

Pig Parts

- Trotters, the hooves of pigs (they have four hoofed toes on each foot, walking mainly on the larger central two).

Biology

- In pig, pregnant.
- Farrowing, giving birth.
- Hogging, a sow when on heat (during estrus).

Housing

- Sty, a small pig-house, usually with an outdoor run or a pig confinement.
- Pig-shed, a larger pig-house.
- Ark, a low semi circular field-shelter for pigs.
- Curtain-barn, a long, open building with curtains on the long sides of the barn. This increases ventilation on hot, humid summer days.

Environmental and Health Impacts

- Industrial pig farming, a subset of CAFOs (concentrated animal feeding operations) poses numerous threats to environmental health and justice. Deforestation is a common occurrence under the placement of new farms. Feces and waste often spread to surrounding neighborhoods, polluting air and water with toxic waste particles. Waste from swine on these farms carry a host of pathogens and bacteria as well as heavy metals. These toxins can leach down through the soil into groundwater, polluting local drinking water supplies. Pathogens can also become airborne, polluting the air and harming individuals when ingested. Contents from waste have been shown to cause many detrimental health implications, as well as harmful algal blooms in surrounding bodies of water.

Fur Farming

Most of the animals used for fur are raised on factory farms. The number of animals killed every year to make fur products remains unknown, though some estimates are between 40-60 million, of which 30-50 million may be raised on fur farms. Though most of the animals used in fur farming are in the European Union, the fur farming industry is growing in China, and other countries such as the U.S. and Canada also have considerable fur farming industries.

To make one fur coat, it takes 150-300 chinchillas, 200-250 squirrels, 50-60 minks, or 15-40 foxes, depending on the animals' subspecies. For the production of fur to be more economical, animals are kept for their whole lives in tiny cages in which they can only move a very small amount and can never do things such as run or swim. This is especially stressful for semi-aquatic animals such as minks, because although they have drinking water, they never have access to most significant sources of it.

Having such little space to live in causes severe stress for the animals, resulting in self-mutilation and cannibalism. Even infanticidal behaviors sometimes occur, with most being cases of mothers eating their young. These are highly abnormal behaviors for minks. Due to confinement and lack of activity, they become frustrated and frequently exhibit stereotyped behaviors, like repetitively moving in a certain way for no apparent reason. In one of many mink farms, a female mink confined in a 75 x 37.5 x 30 cm (30 x 15 x 12 inches) cage was seen repeatedly standing up to grasp the ceiling of the cage and then falling down and onto her back. Similar types of behavior are seen in human beings who, in certain moments, feel a lack of control over some important aspect of their lives, as in situations of deep stress or confinement.

Being caged is a reason in itself for animals to be distressed. So that workers don't have to clean the spaces in which the animals are kept, the floors of their cages are made with wire so the animals' excrement can pass through to pile up below the cages. This means that the cage floors are uncomfortable for these animals. They have to step, sit, and lie on the wire net beneath them for their entire lives. The piled-up excrement is not only a source of possible diseases and parasites, but also a cause of suffering for these animals; the strong stench of excrement is very bothersome to minks due to their acute sense of smell.

These animals also suffer significant discomfort and sometimes pain because of weather conditions. They may have to endure freezing cold in the winter and scorching heat in the summer, and sometimes, as explained in the section on weather conditions, they can die due to heat stress. Also, even though farms are often covered, when there is heavy rain the cold water or snow can still reach them.

Several methods used to kill minks often only leave them unconscious. The most common ones are anal and oral electrocution, neck breaking, and suffocation. The animals are also often skinned alive while they are still conscious.

In the past few decades, there has been a growing social awareness regarding the ethical problems with the use of fur. Therefore, the fur industries have implemented a strategy of including fur in only certain parts of coats, such as necks, sleeves, and hoods. For this reason, in order to avoid financing the fur industry which causes the deaths of all these animals, it is important to be careful when buying coats and jackets.

Minks

Minks are small mammals in the Mustelidae family. Those that are most often used for fur production are American minks. Minks bred by the fur industry commonly spend the majority of their short lives on the same factory farm on which they are born – and eventually killed – without ever going outside.

Minks give birth once a year during the spring. The babies remain with their mothers for several weeks, after which they are then taken away and separated forever. They are killed at about six months of age, usually during November or the beginning of December.

There are several methods used to kill the minks. Farmers gas them with carbon dioxide or sometimes nitrogen. In many cases, in order to reduce production costs, carbon dioxide is used in low concentrations. This causes a slower death. With carbon dioxide concentrations around 70% it can take about 15 minutes of pain before the animals die.

The gases emitted by tractor exhaust pipes have also been widely used. Even though this method has been prohibited in some countries due to the pollutants these gases contain, it is still used. The gases cause stress and convulsions in the animals before they die. Unlike humans and other animals such as pigs, minks are able to detect anoxia (lack of oxygen), which stresses them intensely and causes them much suffering when they are killed. The method of killing minks that is considered "less cruel" is through injections of chloral hydrate or pentobarbital. However, it takes several minutes to kill the minks, and during this time they can feel pain and anguish. This shows that there is not any way to kill minks that is more "humane" than others; each method causes them to suffer. Since chloral hydrate can cause gasping and muscle spasms, pentobarbital injections are preferred by the industry because this allows the mink killers to take the animals back to their cages for them to die. Other methods that are less frequently used are electrocution and neck dislocation.

Rabbits

Rex rabbits are the rabbit breed traditionally used by the fur industry. Babies are kept with their mothers for the first 4-5 weeks of their lives, and then they are put in different cages with their siblings. Finally, when the rabbits are 7-8 weeks old, they are taken from their siblings and spend 1-2 more weeks in a cage completely alone, and then they are killed.

In the mid-1980s the INRA, a French governmental organization, started the Orylag breeding program. The Orylag is a new breed of rabbit that has been bred for commercial purposes. Orylag rabbits are exploited for both meat and fur, with the profit coming mostly from the sale of the fur (60%). Breeding females are made pregnant again by artificial insemination between three and seven days after they first give birth. Rabbits not used for breeding are killed when they are about 20 weeks old.

Rabbits also suffer from confinement in cages. The industry standard for the spaces allowed for rabbits raised for fur, or for fur and flesh, is one rabbit per 60 x 40 x 30 cm (24 x 16 x 12 inches) cage. This is only about as much floor space as two shoe boxes would occupy. In bare wire mesh cages, rabbits are sometimes separated from each other to prevent fighting but are often crammed together. The rabbits can hardly move and may develop bone disorders. Sometimes the rabbits develop deformation of their vertebral column. The cages also prevent the rabbits from sitting up with their ears erect and prevent them from digging, both of which are innate behaviors.

Rabbits are social animals, and being separated from each other is stressful for them. Rabbits that have been separated may develop stereotyped behaviors such as gnawing on the bars of the cage and excessive grooming. Overcrowded housing also causes many problems, and it leads to behaviors such as fur-plucking and ear-biting.

The mesh flooring in cages can lead to sore hocks (ulcerative pododermatitis), which can lead to infections and abscesses. In 2003 it was found that up to 15% of female rabbits suffered from sore hocks, and other research has shown that up to 40% showed discomfort due to paw injuries.

Mortality rates during transport to slaughter can be as high as 7-8%. Broken bones, traumatic lesions, respiratory failure, and the spread of viruses are all common. But many Rex farms carry out slaughter themselves. The rabbits are hit on the head with a club or a tree branch on smaller farms, or stunned by electrocution at larger farms or commercial slaughterhouses. The rabbits are then killed by slitting their throats and letting the blood drain.

Foxes

The foxes most often used by the fur industry are common foxes and arctic foxes. Foxes have been selected because of the desirability of their fur, and also because they are normally docile and bite fur farm workers less frequently. Foxes are normally independent animals who only live in couples or in hierarchical groups during mating and while taking care of their offspring. However, on a fur farm, they spend their lives in tiny cages in which they are surrounded by many other animals in neighboring cages. Foxes develop psychological problems in this environment, showing anxiety, panic, and mistrust; they adopt aggressive and fearful behaviors from being confined. Foxes are only taken out of their cages for farmers to classify them according to their fur, to receive certain veterinary treatments, or if they have to be transferred to another cage for insemination or to be killed.

The way foxes are handled is by holding their necks with 50 cm (28 inch) long steel pliers, with a hole of 7.5 cm (3 inches) diameter for the neck of females and 8.5 cm (3.5 inches) for males. The use of these pliers causes injuries to the foxes' mouths and teeth when they try to escape by biting the metal.

Foxes reproduce once a year. They give birth in the spring, and the offspring stay with the mother for around a month and a half. At this point, the children are weaned and put into separate cages, each of which will be shared by two of them. In November or December, when their fur has developed, the foxes are killed.

Foxes are usually killed by electrocution, using a device consisting of two electrodes with which a discharge is applied. The electrodes are put into their mouths and anuses, and the electric discharge kills them over three to four seconds. Foxes are also killed by injecting pentobarbital or other anesthetics into their hearts.

Chinchillas

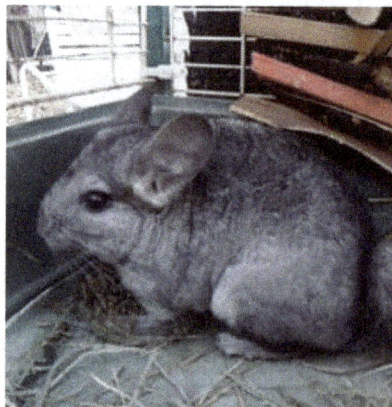

Chinchillas are rodents who have dense fur, which is needed because of the low temperatures in the area to which they are native, the Andes. Some of the countries in which many chinchillas are killed for their fur include Argentina, Brazil, Croatia, the Czech Republic, Poland, and Hungary. However, the main demand for this fur is in Japan, China, Russia, the U.S., Germany, Spain, and Italy.

There are two types of cages in chinchilla farms: breeding cages and growing cages, which commonly hold only one animal. Young chinchillas are separated from their mothers at 60 days of age. Cages may be piled one on top of the other, so that it is possible to have the maximum number of animals in the minimum space. Because of the lack of room, the cage changes, and the separation of the young chinchillas from their families, it is common for them to suffer much distress.

The ways chinchillas are killed include gassing, electrocution, and neck fracture. Electrocution is most common and is used to kill large groups of chinchillas, and neck breaking is used on smaller groups. Electrocution is mainly carried out by applying the electrodes to one ear and to the tail of the animal. There are concerns that these deaths are often painful and that the chinchillas are often not killed immediately. The animal welfare stipulations in place require that heart rate and respiration should be checked to make sure that the animals are dead, but this is often not done. When the chinchillas are killed by breaking their necks, they are held by their tails with their heads hanging down. Their heads are then held and twisted rapidly until the animals die. The pain these animals endure as they are killed adds to the devastation of their being killed unnecessarily in the first place.

Throughout the ages, fur pelts from animals have been traded and worn for their warmth and as a fashion statement. Wild fur-bearing animals have been both trapped and hunted in many countries for ages. In Russia, fur served as a form of currency, was used as gifts and as part of a bride's dowry, and became a significant part of trade during the tenth and eleventh centuries. In the 1530s, the beaver became a main trading item between the American Indians and the colonists, and beaver pelts were regularly shipped to Europe. By the late 1500s, fur was extremely popular in Europe. In 1608, Samuel de Champlain, a French explorer, created a trading post in Quebec, which became the center of fur trade in America. In the seventeenth century, Siberia's unification with Russia helped to propel Russia to become the largest fur-supplier, which it remained until the nineteenth century. Around that time, fur farming started in North America, and was introduced into Europe in the early twentieth century.

A. Facts about Fur Farming

Today, fur farming comprises about 85% of the fur trade, while fur from the wild accounts for the rest. Fur farming is a separate industry from animals that are bred for meat. The meat from some animals bred for their fur, such as rabbits, is sometimes used. However, due to the nature of the intensive confinement systems in which they live, the fur from rabbits bred for their meat is usually not high quality, and thus is often not used. On fur farms, with the help of strict methods of breeding and diet plans, farmers have been able to create "desirable mutations" and high-quality furs. The two most commonly bred animals on fur farms are mink and fox. Other fur-farmed animals include fitches (also known as European polecats, which are related to the ferret), finn raccoons, chinchillas, and nutria. Additionally, goat, sheep, fetal and newborn karakul lambs are sometimes bred for their fur. In 2009, 64.73% of the world's mink and 55.6% of the world's fox fur came from European fur farms. Denmark produced the most farmed mink pelts in the world (30.1%) while China produced 19.3% (the U.S. had the fifth highest amount). The Netherlands (producing the third most, or 9.7% of all farmed minks), Finland, and Sweden are also top producers of fur in the EU, with Finland producing the largest quantity of fox pelts in Europe. Additionally, fur farming is considered a "key industry" for Poland (producing fourth most, or 7.9% of all farmed minks). Latvia, Lithuania, and Estonia are also fur producers as is Kastoria, a Greek town that is one of the main fur manufacturing centers in the world.

North America also plays a significant role in the fur market. Fur farms in North America were the first to breed black mink, which is the most popular mink pelt. In 2009, black mink pelts accounted for 52% of all pelts produced in the United States. Fur Commission USA, which is an association representing 400 mink farmers in the U.S., reports that in the U.S., most fur farms are "family businesses, often operated by two or three generations of the same family." In 2010, the Commission reported having farmer members in Colorado, Connecticut, Idaho, Illinois, Iowa, Indiana, Maine, Maryland, Massachusetts, Michigan, Minnesota, Missouri, Montana, New Hampshire, New Jersey, New York, North Carolina, North Dakota, Ohio, Oklahoma, Oregon, Pennsylvania, South Dakota, Utah, Vermont, Virginia, Washington, West Virginia, and Wisconsin. The U.S. also hosts mink, fox, chinchilla, rabbit, bobcat, lynx, and finn raccoon farming.

The International Fur Trade Federation ("IFTF"), which is comprised of national associations within 35 countries, alleges that it promotes strict codes of practice that "meet or exceed established and accepted standards for animal welfare fur farming is well regulated and operates within the highest standards of care." However, the animal rights organization PETA alleges that animals on fur farms are killed by "anal and vaginal electrocution." Animal advocacy organization Born Free USA reports that animals confined in cages at fur farms suffer "physical and behavior abnormalities" and also may be killed from being gassed, having their necks broken, or being injected with poisons.

B. Facts about Wild Fur

The most commonly traded types of wild fur are North American beaver, coyote, ermine, grey fox, red fox, marten, mink, muskrat, nutria, opossum, raccoon, Russian sable, and Chinese weasel. Most wild fur is obtained from Canada, Russia, and the U.S. Trappers generally use "baited and concealed traps" during the time of year when the desired animal has the highest quality coat (this is usually at the start of winter). Wild fur is mainly traded by hunters and trappers in U.S., Canadian, European, and Russian auction houses, where pelts are bought and subsequently processed into clothing.

The IFTF states that the majority of wild fur is obtained from wildlife management programs, which are necessary for "maintenance of biodiversity and healthy eco-systems, population and disease control and the protection of public lands and private property." Additionally, the Association of Fish and Wildlife Agencies alleges that trapping is beneficial to conservation efforts. However, animal welfare organizations say that trapping has many harmful effects on wildlife. First, they argue that trapping can cause species to become endangered once popularity of pelts rise and those species become more prone to extinction. Second, they allege that trapping can increase the spread of disease, as healthy animals are more likely to be lured into traps than weak animals, thus reducing the "genetic strength" of the animals. Third, they argue that trapping can contribute to the overpopulation of wildlife, as it can cause some species to reproduce more quickly than normal, which can upset the "delicate and complex balances that exist in nature." PETA reports that animals in the wild are drowned, trapped, or beaten to death. Born Free USA reports that animals caught in traps may remain there "for several days before starving or dying from exposure." Snares can slowly strangle animals to death, and animals whose limbs are caught in leg-hold traps may chew off their paws to escape and then subsequently die from the injury. Additionally, traps can be dangerous because they can kill or injure any animal that comes into contact with them. Animals that are trapped by mistake are deemed "trash animals" because they do not have any economic value to the trappers, and according to Global Action Network, an animal and environmental protection organization, "trappers themselves report that three to ten 'nontarget' animals are caught in the trap for each intended victim."

C. Other Parts of the Fur Trade

Some aquatic animals such as harp seals are hunted for their pelts. Canada and Greenland's commercial seal hunts make up 50% of all seals hunted, while Namibia hosts the only hunt south of the equator. Baby harp seals in Canada cannot be legally hunted until they begin to shed their white coats of fur. However, this happens at a very young age. In 2006, the Humane Society International, an animal protection organization, reported that "97 percent of the seals killed in the commercial seal hunt over the past three years have been younger than 3 months, and most were younger than 1 month old." Polar bears are hunted for their fur and meat by indigenous communities in the Arctic, but they are also used as trophy pelts. Additionally, dogs and cats are part of the fur trade. Global Action Network alleges that 2 million dogs and cats are killed in primarily China and South East Asia every year for the fur trade, and that dogs are traditionally killed by being hung from their paws and are bled to death from an artery cut in their thigh, while cats are often "strangled with wire nooses."

I. Animal Rights Organizations and the Future of Fur

In 1988, the World Society for the Protection of Animals, an alliance of animal welfare organizations, launched its "No Fur" Campaign, which was subsequently adopted by over fifty other organizations. In the same year, PETA launched an undercover investigation of a Montana beaver farm, which led to its closure, and started to use donated furs in demonstrations and educational displays. Around the same time, PETA also made headlines when one of its members put a dead raccoon onto Editor-in-chief of fashion magazine *Vogue* Anna Wintour's plate at the Four Seasons, and hit fashion designer Oscar de la Renta in the face with a tofu cream pie. Anti-fur campaigns successfully created a negative image of fur, and activist protests led to a decline of fur used in fashion during the 1980s; however, fur has been making a comeback since the 1990s, and in March 2010, "for the first time in more than two decades, more designers are using fur than not."

II. Production and Fur Sales

In 2008, about 56 million pelts of fur were produced worldwide. Fur trim is a large market and the number of animals killed for it is predicted to exceed those killed for clothing made entirely by fur. The U.S. is a "major exporter of pelts," but its fur apparel production has decreased significantly, thus making it a net importer. The U.S. is able to acquire some of the highest prices for its mink pelts, but now it has less than 300 farms, mainly in Wisconsin. In 2008, the U.S. imported fur mainly from China (44.4%), Canada (16.4%) and Italy (13.9%), and exported fur to China (21.5%), Canada (13.4%) and Italy (8.2%). In March 2010, the IFTF reported that global fur sales remained stable during the 2008-2009 global recession, and while fur sales in Europe have remained around $4 billion for 2007-2009, Asian markets have experienced increased demand, for a total global fur sale figure "at just over $13 billion." This number is stable from the past few years, but lower than the global $15.02 billion in 2006-2007.

Russia and China are increasing their production of and market for fur, and China's cheap labor and lack of regulatory oversight have attracted more business. It is difficult to get a true sense of China's market share, since most of its fur farms are family-owned businesses, the pelts are often "sold on the free market" (as opposed to auctions), and the country lacks official statistics about its fur production, yet at the same time the U.S. Department of Agriculture ("USDA") believes that China is the world's leading fur processor.

A. Real Fur Sales Versus Faux Fur Sales

For people who like the look of fur, but do not want to kill animals, faux fur can be an attractive alternative. Faux fur is usually made of acrylic fibers that can be dyed to look like animal fur. There has been recent controversy over real fur which is unlabeled or mislabeled as faux fur. Due to this fact, HSUS has created a guide to help people to determine whether a product is made of real or faux fur:

1. Check the base of the fur for skin or fabric. Push apart the fur and look at the material at the base of the hairs. If the base material is not visible or unclear, and you own the garment, break the stitching and look at the non-hair side of the fur base, being sure to peel away all the layers of the lining.

Animal Fur: The surest sign of animal fur is leather/skin (usually white or tan, but possibly the color of the fur if it has been dyed).

Fake Fur: The surest sign of fake fur is seeing the threadwork backing from which the "hairs" emerge.

2. Check the tips of the hairs for tapering. Both animal fur and fake fur come in many different colors and lengths. However, if animal fur has not been sheared or cut to a uniform length or had the guard hairs plucked out, you may be able to examine the tips of the longest hairs and see that they taper into a fine point—like a cat's whisker or sewing needle. Good lighting and a magnifying glass are helpful, as is holding the hairs up against a white surface.

Animal Fur: Animal hairs—especially the thicker guard hairs —can often be seen tapering to a point.

Note: This test can give a false negative for animal fur if the hairs have been sheared or plucked.

Fake Fur: There is a straight across cut in fake fur "hair."

Note: Tapering has not been seen on any fake fur samples to date, but such a process may exist, or come into existence.

3. The Burn Test: Animal hair smells like human hair when burned; fake fur made from acrylic or polyester—the two most commonly used synthetics—does not. Carefully remove just a few hairs and then, holding them with tweezers above a dish or other non-flammable surface, ignite them with a cigarette lighter. Make sure to burn them away from the original garment and anything else flammable. Never conduct the burn test on hairs still attached to the jacket. The burn test should only be conducted by adults.

There is controversy over whether faux fur or real fur is more "environmentally friendly." A 2004 British newspaper article quotes Executive Director of Fur Commission USA Teresa Platt as saying that "faux fur jackets do not degrade for at least 600 years and may take thousands of years," and Ruth Rosselson from *In Touch* magazine as saying that polyester and nylon, which are used to make fake fur, are synthetic materials that are "responsible for large-scale factory pollution of our waterways, rivers, canals and even the sea." However, PETA alleges that it takes twenty times more energy to produce a real fur coat from farm-raised animals than it does to create a faux fur coat, that fur is not biodegradable because it contains chemical treatments to prevent it from rotting, and that "mink factory farms generate tens of thousands of tons of manure annually and can produce nearly 1,000 tons of phosphorus, which wreaks havoc on water ecosystems."

III. US Federal Regulation of the Fur Industry

There is very little federal law regarding treatment of fur animals. While U.S. Congress has created the Animal Welfare Act to ensure humane treatment of animals, it specifically exempts "animals used or intended for use as food or fiber." Similarly, the Humane Methods of Slaughter Act , which requires livestock to be slaughtered humanely to prevent "needless suffering," does not extend protection to fur animals. An examination of all the federal laws that reach fur animals is instructive.

A. The Lacey Act

The Lacey Act, 16 U.S.C. 3371-3378, prohibits wildlife trade in animals that have been illegally taken, possessed, transported or sold. However, this law cannot always prevent poaching because "state laws vary widely. If a poacher kills a bear in a state that prohibits trade in bear parts, the poacher can avoid prosecution by transporting the body to a state that does permit it. Although such transporting is illegal, a prosecutor must prove that the bear was illegally killed in a state that prohibits commerce in bear parts, which can be very difficult to do." The Lacey Act only applies to wild fur and not to fur obtained from fur farms.

B. The Fur Seal Act

The Fur Seal Act, 16 U.S.C. 1151-1187, makes it illegal to take North Pacific fur seals anywhere in the U.S., and mandates the Secretary of Commerce to regulate the fur seal breeding colonies on the Pribilof Islands, (part of Alaska), to ensure that activities on the Islands do not deter the conservation of the North Pacific fur seals. However, Indians, Aleuts and Eskimos who live on the North Pacific coast are allowed to take fur skins if they are taken for subsistence. Additionally, the Secretary of Commerce is allowed to take North Pacific fur seals or parts of these seals if deemed for education, science, or for an exhibition. The Fur Seal Act only applies to wild fur and not to fur obtained from fur farms.

C. The Marine Mammal Protection Act

The Marine Mammal Protection Act, ("MMPA") 16 U.S.C. 1361 - 1421h, was created to protect marine mammals that are in danger of extinction or depletion because of human activities; the Act applies to mammals that primarily live in the water and to all parts of the mammal, including its fur. The MMPA prohibits the taking of marine mammals, except in the case of permits issued by the Secretary of Interior for "scientific research, public display, photography for educational or commercial purposes, or enhancing the survival or recovery of a species or stock or for importation of polar bear parts (other than internal organs) taken in sports hunts in Canada," as well as for commercial fishing operations (when marine mammals are taken incidentally). The Secretary can also waive the requirements of the law for specific circumstances. The MMPA only applies to wild fur, and not to fur obtained from fur farms.

D. Polar Bear Bills Pending in Congress

Currently, there are three bills in Congress addressing the taking of polar bears from sports hunts in Canada. H.R. 1054, introduced by Representative Young from Alaska, would have the Secretary issue a permit allowing polar bear parts (besides internal organs) to be imported from Canadian sports hunts that occurred before the polar bear was determined to be "threatened" under the Endangered Species Act. The bill was referred to the Subcommittee on Insular Affairs, Oceans and Wildlife, subcommittee hearings were held in 2009, and no further action was taken. Sen. 1395, introduced in 2009 by Senator Crapo from Idaho, contains very similar language. It was referred to the Committee on Commerce, Science, and Transportation, and no further action was taken. H.R. 1055, also introduced by Representative Young from Alaska, would allow polar bear parts to be imported from all Canadian sports hunts (thus, it is broader than H.R. 1054).   The bill was also referred to the Subcommittee on Insular Affairs, Oceans and Wildlife in 2009, and no further action was taken.

E. Fur Labeling

The Fur Products Labeling Act, 15 U.S.C. 69, declares that fur products will be considered "misbranded" if "falsely or deceptively labeled" or identified, and/or if the product does not contain a label that legibly shows the names of the animals from which the fur was taken, the name or other identification of the persons who manufactured the fur, and the country of origin of the fur. The label must also state, if true, that the fur product contains used or artificially colored fur, and/or if it is "composed in whole or in substantial part of paws, tails, bellies, or waste fur." However, the law defines "fur product" as an article of clothing that is made in whole or in part by fur, but states that the Commission can exempt articles because of the small quantity of fur they contain. The Federal Trade Commission has deemed "relatively small quantity or value" to equal $150, which means "multiple animal pelts can exist on a garment without a label."

HSUS has been advocating for an amendment to the law closing this loophole, after its investigators found "dozens" of designers selling unlabeled jackets that contained animal fur, and mislabeled jackets that said they contained faux fur, but really contained animal fur. In May 2009, Representative Moran introduced H.R. 2480, the Truth in Fur Labeling Act, which would amend the Fur Products Labeling Act by eliminating the exemption for fur products under $150 that allows them to be unlabeled. This act would not apply to fur from hunting or trapping which is sold by the hunter or trapper "in a face to face transaction," where the money made is not the person's primary source of income. On July 28, 2010, the bill passed the House, and was referred to the Senate Committee on Commerce, Science, and Transportation on August 5, 2010. Supporters of the bill think that consumers will be better protected by this legislation because they will be able to make "informed purchasing decisions," and they allege that companies will no longer be able to take advantage of the $150 loophole (since the loophole allows cat and dog fur to "slip into the United States despite the ban.") Detractors of the bill compare fur farming killing methods to those used by animal shelters, they say that cat and dog fur are not traded in North America, and that all fur meets appropriate labeling requirements and is farmed legally.

F. Dog and Cat Fur Protection Act

The Dog and Cat Fur Protection Act, 19 U.S.C. 1308, prohibits the import, export, and sale of dog and cat fur products in the U.S. The law requires that the Secretary of Treasury submit a report to Congress every year on the government's enforcement efforts and its ability to do so. The report is to include any findings that a particular government has supported the dog or cat fur trade.

G. Endangered Species

The Endangered Species Act, 16 U.S.C. 1531-1541, works to develop conservation programs to protect endangered and threatened species, acknowledges the U.S.' commitment to conserving, as much as practical, wildlife under the Convention on International Trade in Endangered Species of Wild Fauna and Flora ("CITES"), and encourages states to create conservation programs (with federal financial assistance) that meet national and international conservation standards. A species is deemed "endangered" if it is "in danger of extinction throughout all or a significant portion of its range" or "threatened" if it is "likely to become an endangered species within the foreseeable future throughout all or a significant portion of its range." Fur-bearing animals on the list of endangered species include some types of bears, beavers, cheetahs, leopards, monkeys, rabbits, tigers, yaks, and zebras. Although the Endangered Species Act protects animals on its list, it allows people to own endangered species, and may even allow them to hunt those animals. Thus, some species that are listed may in fact be hunted for their fur.

IV. US State Regulation of the Fur Industry

It is often times unclear under what classification of laws, fur animals belong. Trapping and hunting laws are generally under State Wildlife, Fish & Game, or Environmental Conservation Codes, labeling laws are under Trade Practice Codes and cat and dog fur laws are under anything from the Criminal Code to the Agriculture Code.

Fur Commission USA reports that state departments of agriculture regulate fur farms. However, specific laws regarding such regulation are scarce, and are mainly confined to defining fur farming as an "agricultural pursuit" and designating the animals as either domestic animals or livestock. For example, Wisconsin, which has the most fur farms in the U.S., has a law stating that fur farming is an "agricultural pursuit, and all such animals so raised in captivity shall be considered domestic animals," and thus subject to the laws for domestic animals. Michigan has a similar law, which declares "silver, silver-black, black, and cross foxes" on fur farms to be domestic animals, as distinguishable from livestock. In California, animals that are fur farmed are considered to be domestic animals, yet the statute differentiates between fur animals and "dogs and cats or other pets," and states that the animals will be considered domestic animals for the purposes of laws relating to farming and animal husbandry. In Idaho, fur farming is again defined as an "agricultural pursuit," but here, the animals are considered to be livestock. Idaho has some of the most extensive laws, including one giving its Animal Industries Division the right to inspect fur farms at any time. Minnesota requires a fur farmer to register with the state to provide information about the number of pelts he or she sells. In Montana, a farmer must obtain a fur farm license in order to operate a fur farm, but the law specifically excludes fox and mink fur farms.

Born Free USA published a 2009 report which alleged that many states allow fur farms to be unregulated:

No states reported having comprehensive laws specific to the regulation of fur farms and no states monitor the care and treatment of animals housed and killed on fur farms. As a result, fur farms are virtually unregulated in every state where fur farming exists.

In response to formal requests for information from Born Free USA, the vast majority of Departments of Agriculture in fur-farming states reported having no specific responsibilities or regulatory authority over fur farming in the state. Of those states reporting that their Department of Agriculture has statutory authority to regulate fur farms (Idaho, Massachusetts, Michigan, Minnesota, New York, South Dakota), none had exercised this authority by issuing regulations.

A. Trapping Laws

Born Free USA alleges that trapping laws are state-regulated and are often poorly enforced. Trapping laws typically specify the season during which species can be taken. They also set up licensing procedures, with administrative regulations detailing the types of traps that hunters can use. Many states do not restrict the type of traps, how many animals can be trapped, and the regularity in which trappers must check their traps. However, a significant amount of states have banned leg-hold traps, which is a trap designed to hold an animal's limb so the rest of its fur is not damaged. Colorado, New Jersey, Washington, and Massachusetts are such states. In Arizona, leg-hold and instant kill traps are banned on public lands. In Rhode Island, leg-hold traps are banned, but a person can apply for a special permit to use a leg-hold trap on his or her property when he or she has an "animal nuisance" that "cannot be reasonably abated except by the use of the trap." In Maryland, leg-hold traps cannot be within 150 yards of a "permanent human residence." In other states, such as California, New York, and Iowa leg-hold traps of a certain diameter are banned. Wyoming permits leg-hold traps, but requires that the trapper check them at least once every 72 hours.

B. Fur Farming and Trapping under Anti-Cruelty Laws

Lines are also somewhat vague in terms of animal cruelty laws and how they relate to wild fur animals and those raised on fur farms. While state anti-cruelty laws generally exclude legal trapping or agricultural pursuits, sometimes this distinction is not so clear. In Florida, a man was convicted of felony cruelty to animals after he shot an opossum with a BB gun that he had found in his garage, shooting it so many times and injuring it so severely that it had to be euthanized. Even though the opossum was a "fur-bearing animal," which meant that a licensed hunter could "take" it, the relevant cruelty law "applies to even the unintended consequence of a lawful act." In fact, the court observed that this was a "blurred line between the lawful hunting of an animal and the commission of a criminal act."

Alabama's anti-cruelty law applies to the cruel mistreatment of any animal. However, most other anti-cruelty laws exempt animals hunted or raised on fur farms. Arizona's anti-cruelty statute does not "prohibit or restrict" any activities that are deemed to be regulated by Arizona's Game and Fish Department or its Department of Agriculture. Similarly, Idaho forbids a person from treating any animal cruelly, but further states that the law is not meant to interfere with "normal or accepted practices of animal identification and animal husbandry" or "any other practices or procedures normally or commonly considered acceptable." Connecticut and Texas have similar laws, but exempt hunting and "generally accepted" agricultural activities. While state cruelty laws do not generally apply to either trapping or fur farming activities, some states have enacted companion laws that require humane euthanasia for fur-bearing animals. For example, New York specifically prohibits electrocution of fur-bearing animals, while California prohibits using carbon monoxide gas or an injection directly into the heart muscles or ventricles of a euthanasia agent to kill a conscious animal, unless the animal is "heavily sedated or anesthetized in a humane manner."

C. Labeling Laws

Five states have labeling laws that are stricter than the federal fur labeling law. Delaware and New Jersey require garments containing real animal fur to contain labeling stating so. Wisconsin has a similar law, but the labeling requirement does not apply to garments that are sold for less than $50. In Massachusetts and New York, both real and faux furs must be labeled as such. In California, A.B.1656 passed the Assembly and Senate in August 2010, which would require labels on all clothing that contained animal fur, but the bill still needs to be approved by Governor Schwarzenegger, who has not yet disclosed his position.

D. Dog and Cat Fur

A handful of states have laws concerning dog and cat fur. Alabama, Delaware, New Jersey, New York, Pennsylvania, and Virginia prohibit trade in domestic dog or cat fur. Virginia also prohibits killing a dog or cat for its fur, while Florida prohibits killing a dog or cat with the "sole" intention of either selling or giving away the pelt of the animal. In Oregon, a person cannot buy or sell dog or cat fur that is obtained from "a process that kills or maims the cat or dog."

Although fur farming makes up 85% of the total fur production, there are very few on-point laws regulating it. While there are a plethora of laws regarding trapping licenses, and a few laws banning certain types of traps, states do not have laws regulating fur farms, outside of the need to obtain licenses and the definition of the fur animals. Additionally, only a handful of states have fur labeling laws and prohibit dog and cat fur trade, although there are federal on-point laws for each of these subjects.

V. Commerce in Illegal Fur

Even though the majority of fur obtained worldwide is from fur farms, an illegal wildlife trade still flourishes. This trade, (including the illegal fur trade) is annually worth between $5-$20 billion, with the Congressional Research Service reporting that "'the illegal wildlife trade is among the most lucrative illicit economies in the world behind illegal drugs and possibly human trafficking and arms trafficking.'" Much of the illegal fur is obtained from endangered animals that are against the law to hunt.

A. Tiger and Leopard Fur Trade

Tigers and leopards are both considered endangered species, and thus their international commercial trade is prohibited under CITES. According to a report conducted by the Environmental Investigation Agency ("EIA") and the Wildlife Protection Society of India ("WPSI") in 2006, "the illegal trade in poached skins between India, Nepal and China is the most significant immediate threat to the continued existence of the tiger in the wild." Although India and China are signatories to CITES, which has given recommendations to help stop illegal trade, the countries have not implemented the recommendations and leopard and tiger skins have been subject to trafficking by international organized crime networks. Poachers can be paid $1,500 in India and $16,000 in China for one tiger skin; fines in India and Nepal are only $440 and $1,420, respectively.

Tiger and leopard skins originally were used mostly for Tibetan chupas, which are worn at festivals and weddings. However, now there is an increasing demand for the skins to be used as decorations in the home and expensive gifts. Tigers are caught in traps and then shot, or clubbed or speared through the mouth so that the skin remains intact. Professional tiger and leopard poachers from traditional hunting communities have become famous for using steel jaw traps and guns, and also for poisoning and electrocuting the animals.

In 1993, China illegalized imports, exports, transport, and purchase of tiger products. However, at a 2007 CITES meeting, China said that the trade ban had proven ineffective because wild tiger populations were still decreasing, and the ban had "seriously impacted the Chinese traditional culture." China argued that it should be legal to farm tigers for their bones and skin, and the country has allowed tiger breeding farms. In 2007, China reported having 5,000 tigers on its farms.

At the March 2010 CITES meeting, all member nations agreed to increase law enforcement and cooperation between countries where tigers exist, as well as to "create a tiger trade database that could be analyzed to develop anti-poaching strategies." The South Asia Experts Group on Illegal Wildlife Trade will work to respond to poaching and trafficking issues. Additionally, a potential resolution was introduced at the 2010 CITES meeting to create further limitations on the domestic trade of tigers, but it did not pass, and regulations of tiger farms were not modified. However, HSUS reports that there was a "renewed call" for countries such as China who allow tigers to be bred for commercial trade, to end such activity.

B. Polar Bear and Bobcat Fur Trade

At the 2010 CITES meeting, the U.S. proposed to change the polar bear's listing status on the CITES appendices; this was defeated by the European Union's ("EU") opposition to such a change. Such a ban would have moved the polar bear from inclusion on CITES' Appendix II, where regulated international commercial trade is permitted, to inclusion on Appendix I as an "endangered species," where no international commercial trade is allowed.

The U.S. also attempted to remove the bobcat from Appendix II, where its skins can only be exported if the exporting country determines that the export will not hurt the species' chances of survival. However, this proposal was defeated.

VI. Other International Trade Issues

International countries have a wide range of laws pertaining to fur-bearing animals. For an example, China has virtually no regulations protecting such animals. On the other side of the spectrum, the EU has a host of regulations. Animals kept for farming purposes, including fur-bearing animals, are protected under EU Council Directive 98/58, which mandates, (through the enactment of laws in its member states), that animal owners "ensure the welfare of animals under their care and ensure that those animals are not caused any unnecessary pain, suffering or injury." Additionally, EU Council Directive 93/119 protects animals at the time of slaughter or killing, and includes specific methods in which an animal produced for its fur can be killed.

A. Fur Farming Limitations

Austria and the United Kingdom have banned fur farming, and Croatia began a ten-year phase-out in 2007. The Netherlands has banned fox and chinchilla farming, and strict regulations regarding farming and imports have effectively banned mink farms in New Zealand, fox farms in Sweden, and all fur farms in Switzerland.

B. Leg-Hold Trap Bans

Over 60 countries have banned the leg-hold trap. The EU passed Regulation 3254/91, which prohibits leg-hold traps, as well as fur imports from other countries that were obtained from leg-hold traps or other traps that do not meet "international humane trapping standards." The EU reached and ratified an agreement regarding international humane trapping standards with Russia and Canada, and also reached a similar agreement with the U.S. The Commission also proposed a Directive in 2004 which would introduce humane trapping standards for specific species. The European Economic and Social Committee reviewed the proposed Directive, and found that the Agreement's use of the word "humane" to define the type of trapping allowed, was "controversial, given that these standards are based on the acceptance of high level of suffering for the trapped animals," and therefore recommended that the word "humane" be eliminated from the text, that the drowning trap should be prohibited and a "transparent system" of licensing trappers should be implemented, in order to ensure that appropriate welfare standards are upheld. The proposed Directive has not yet been adopted.

C. Imported Seal Products Ban

In 1983, the EU banned the import of seal pup products made from whitecoat harp seal pups seals and of hooded seal pups (the Directive does not apply to products from Inuit traditional hunting). The U.S., Belgium, Netherlands, Mexico, Slovenia, and Croatia have prohibited the seal product trade. In May 2009, the EU set forth a regulation which banned seal product trade within its countries. Canada and Norway (who is not a member of the EU) have alleged that the ban is a violation of WTO trading obligations. Supporters of the ban say that seal slaughter is cruel and inhumane, while those in opposition disagree, and further say that a ban will destroy the hunters' jobs. The ban started on August 20, 2010, although some hunters are currently exempt, due to a pending lawsuit.

D. Fur Products Ban

On September 2, 2010, Israel is planning to vote on a bill which would "prohibit the production, processing, import, export, and sale of fur from all animal species not already part of the meat industry." The bill provides a "cultural" exception so that Hasidim Jews can still obtain their fur-covered hats (or shtreimels). The IFTF believes that there is no justification for the bill, as it says that the fur trade is mindful of animal welfare and conservation.

E. Cat and Dog Fur Trade and Labeling Laws

The EU banned the import, export, and placing into the market of dog and cat fur in 2007. The U.S. and Australia also have such bans. The IFTF supports these bans.

In terms of labeling, the IFTF reports that the U.S. Canada, Russia, the Ukraine, China, and parts of the EU have fur labeling requirements.

F. CITES

CITES is an international agreement that works to create a sustainable wildlife trade. There are currently 175 parties to the Convention, including China, India, the U.S., the Russian Federation, and Canada. The EU is not a member, but has implemented the agreement's provisions in its community law. While CITES parties are legally bound to the Convention's terms, each party must adopt domestic legislation in order to implement CITES in its particular country. CITES looks at the international trade of certain species and protects them in three different ways, depending under which Appendix a species falls. Appendix I contains species that are threatened with extinction. CITES permits trade of those species only in rare situations. Appendix II contains species that do not necessarily face extinction, but must have controlled trading circumstances in order to protect them. Appendix III contains species that are protected by at least one country that has asked CITES to help it control the trade of that species. Parties are able to submit proposals at regular Conference of the Parties meetings to try to amend either Appendix I or II. All import and export activity of any species falling within an Appendix must be subjected to the Convention's licensing system; for example, a species under Appendix I will only be issued an import permit if it is not going to be used mainly for commercial purposes and if the import's purpose will not hurt the species' survival. The fur trade is directly affected by CITES because, depending on a species' listing under Appendix I or II, it is either prohibited from commercial trade, or its trade is monitored.

Conclusion

The fur industry is predominantly supported by fur farming. It seems, due to the ease in regulating an animal's reproduction and coat quality through farming that this method will continue to predominate over hunting. China's increasing market share and its lack of regulatory oversight (resulting in alleged cruel slaughter methods and the use of dog and cat fur), have the potential to change the fur industry landscape. Some U.S. states have laws concerning the proper labeling of fur and the prohibition of dog and cat fur. The U.S. also has on-point federal laws for both issues, and the EU and Australia have similar laws regarding the sale of dog and cat fur. There are also international movements to protect endangered animals from the illegal fur trade. The U.S. supports such a movement through its role in CITES; however, its pending legislation at home regulating the taking of certain animals who are at least deemed "threatened," (such as polar bears), seems largely dependent upon the political party in office. Additionally, sophisticated fur industry trade organizations have denounced unregulated and illegal fur trades. These associations, when faced with the possibility of losing market share to China, could put pressure on China to devise regulations (and to thus "level the playing field").

At the same time, the U.S. is lagging behind the rest of the world in terms of regulations regarding the treatment of animals that are farmed or hunted for their fur. Many countries have strict rules regulating or banning fur farming. However, the fact that China is the major player in the industry, and also has an increasing demand for fur, begs the question of whether these attempts by other countries are futile, as the fur production is merely "outsourced" to Asia. While the U.S. used to be a big fur producer, this is no longer the case, thus its lack of regulations could soon become moot.

One country, Israel, has introduced a possible ban on the production and sale of fur within its national confines. If this becomes law, its potential to influence other countries – especially ones who have already banned certain fur products – is significant. At the same time, designers have renewed their use of fur on the runway, thus showing fur's continuing prominence in fashion. In all, the future of the fur industry and its prominence in society is very uncertain. Whether China's increasing production or laws calling for more regulation will win, remains to be seen.

References

- "The famous dip the helped cure the scourge of Sheep". Dacorum Heritage Trust. Archived from the original on 3 March 2016. Retrieved 18 March 2016

- Nicole, Wendee (2017-04-21). "CAFOs and Environmental Justice: The Case of North Carolina". Environmental Health Perspectives. 121 (6): a182–a189. doi:10.1289/ehp.121-a182. PMC 3672924. PMID 23732659

- Qushim, B., Gillespie, J.M. and McMillin, K. (2016). "Analyzing the costs and returns of US meat goat farms". Journal of the ASFMRA

- Hemsworth, P.H (2003). "Human–animal interactions in livestock production". Applied Animal Behaviour Science. 81 (3): 185–98. doi:10.1016/S0168-1591(02)00280-0

- Wong, E., Nixon, L.N. and Johnson, B.C. (1975). "The contribution of 4-methyloctanoic (hircinoic) acid to mutton and goat meat flavor". New Zealand Journal of Agricultural Research. 18: 261–266

- McVeigh, T. (March 15, 2015). "Honey, I cooked the kid. How Britain fell for goat meat". The Guardian. Retrieved August 7,2016

- Hemsworth, P.H.; Price, E.O.; Borgwardt, R. (1996). "Behavioural responses of domestic pigs and cattle to humans and novel stimuli". Applied Animal Behaviour Science. 50 (1): 43–56. doi:10.1016/0168-1591(96)01067-2

Permissions

Index

www.ingramcontent.com/pod-product-compliance
Lightning Source LLC
Chambersburg PA
CBHW062000190326
41458CB00009B/2926